NEUROWAVES

NEUROWAVES

Brain, Time, and Consciousness

GEORG NORTHOFF

McGill-Queen's University Press
Montreal & Kingston • London • Chicago

© McGill-Queen's University Press 2023

ISBN 978-0-2280-1761-5 (cloth)
ISBN 978-0-2280-1817-9 (ePDF)
ISBN 978-0-2280-1818-6 (ePUB)

Legal deposit second quarter 2023
Bibliothèque nationale du Québec

Printed in Canada on acid-free paper that is 100% ancient forest free (100% post-consumer recycled), processed chlorine free

 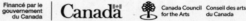

We acknowledge the support of the Canada Council for the Arts.
Nous remercions le Conseil des arts du Canada de son soutien.

Library and Archives Canada Cataloguing in Publication

Title: Neurowaves : brain, time, and consciousness / Georg Northoff.
Names: Northoff, Georg, author.
Description: Includes bibliographical references and index.
Identifiers: Canadiana (print) 20220491666 | Canadiana (ebook) 2022049195X | ISBN 9780228017615 (cloth) | ISBN 9780228018179 (ePDF) | ISBN 9780228018186 (ePUB)
Subjects: LCSH: Time perception. | LCSH: Brain. | LCSH: Consciousness. | LCSH: Mind and body.
Classification: LCC QP445 .N67 2023 | DDC 612.8/2—dc23

Contents

Introduction: The Brain's Mental Surfing 3

1
Brain Time 12

2
From Brain Time to World Time 26

3
The Tango of Brain Time and Body Time 42

4
Self Time and Its Duration 58

5
Time Speed in Brain and Mind 74

6
Beyond Human Time 86

Coda: Copernican Revolution in Neuroscience and Philosophy 102

References 107

Index 117

NEUROWAVES

Introduction:
The Brain's Mental Surfing

What is time? We say that time is short if we need to rush to the supermarket before it closes. While time can stretch and feel endless when we are sitting in a rather boring meeting. Then there is the time of day and night, the twenty-four-hour rhythm of our lives. Historians look far beyond when thinking in terms of hundreds or thousands of years. Biologists and especially evolutionary biologists speak of several thousand if not millions of years. Time is everywhere and shapes our world as well as our mind and its behavior. But what is time itself? This is one of the most fundamental questions that generations of scholars have raised, especially philosophers and physicists in ancient and present times.

Time in Ancient Greece: Chronos and Kairos

The ancient Greeks distinguished between two Gods of time: Chronos and Kairos. Chronos describes chronological or sequential time as characterized by the continuum of past, present, and future. Kairos refers to a particular point or moment in time that signifies the right moment for action. Unlike the continuously changing time of Chronos, Kairos's time is not subject to change but is more permanent.

Chronos and Kairos were brothers or father and son in the ancient myth. That leaves open the relationship between their two senses of time: how is the past-present-future continuum of time, signified by Chronos, related to the proper moment in time, signified by Kairos? Another famous ancient Greek, this time a real person, was the medical doctor Hippocrates, who said that "every kairos is a chronos, but not every chronos is a kairos." The right moment in time, the kairos, is nothing but a specific manifestation of the continuum of time, the chronos – the single moment is one point in time on the continuum of time. However, the reverse does not hold; the chronos as continuum of time is not based on the kairos.

The ancient Greek distinction between chronos and kairos has shaped our modern view of time, especially in the Western world. A moment in time is not extended, while a continuum of time is extended – something that is not extended stands opposite to that which is extended. Continuum and moment are thus conceived to refer to mutually exclusive features of time. This is, for instance, well apparent in the way physics conceives of time.

Time in Classical Physics: The Container View

Classical physics (Newton, Kepler, Galileo, etc.) views time as "container" or "theater" within which events or objects like the brain are contained and located at specific points ("events occur in time," "the brain is in time"). This amounts to what the philosopher Barry Dainton describes as "container view" of time (Dainton 2010, 2–3). Time is like a container that provides the continuum for the "right moments" in time at which the distinct events or objects including the brain occur.

Imagine a trash can and trash. Trash can and trash are separate as they remain independent of each other. A trash can is able to hold something other than the trash, and trash is able to sit somewhere other than in the trash can. Now compare the continuum of time, provided by Chronos, to the trash can and the right moment in time, provided by Kairos, to the trash. Since trash can and trash remain more or less separate, the two ancient forms of time (as well as their respective Gods) remain distinct and independent. I therefore characterize such view as the container view of time.

Time in Contemporary Physics: The Construction View

Contemporary physics does not follow classical physics in its view of time. Contemporary physicists like Lee Smolin (2015; see also Weinert 2013 and Rovelli 2018) no longer make such a stark distinction between continuum and moments in time. Here, time supposedly consists of the continuous construction of spatiotemporal relations between the different objects or events. This renders time intrinsically dynamic, that is, it is continuously changing over time rather than remaining static like a container (i.e., continuum of time) containing its contents (i.e., moments of time).

This amounts to a "construction view" of time as it highlights the continuous construction of spatiotemporal relations. This view of time entails past and present Western philosophers like Leibniz, Bergson, Husserl, Heidegger, and Whitehead describe as "relationism." Such relationism connects the continuum of time to the right moments of time in a way in more or less the same way Chronos was supposed to be the father or older brother of Kairos.

What is relationism? Relationism understands time as a continuous construction of different scales of time, including both short moments that change, and long stretches of continuity with sameness across different time scales. Shorter and longer time scales integrate and relate with each other – that amounts to co-occurrence of both change and continuity. Since time is constructed rather than taken as a given, this amounts to a construction view of time.

Lessons from Ancient China: From Brain Dynamic to Mind Dynamic

The construction view of time is often neglected or misunderstood in the Western world but is much more prevalent in the Eastern world. Anticipating contemporary physics, early ancient Taoist Chinese philosophers like Zhuangzi vividly describe the dynamics of time and its connection between continuum and moments. Chronos and Kairos are here not two different gods of time, but rather one and the same wave, the wave of time featured by its ongoing dynamic.

The dynamics of time must be imagined like waves on the ocean. The ocean never stands completely still, with always something going on. We can observe powerful, slow waves that sweep across large segments of the ocean – this is the continuum of time to which Chronos refers. We can also observe less powerful, fast waves that extend over shorter distances – the moments in time to which Kairos refers.

Both slow and fast waves arise from one and the same underlying ground: the dynamics of water of the ocean. Now to step from time to brain: what the dynamic of the ocean's water is for its waves, the brain's inner dynamic is for the mind. Mental features are themselves constructed in a dynamic way as they are based on the brain's construction of its own inner time relative to the world's outer time. Accordingly, through its own dynamic, the brain's inner time yields our mind. The mind is thus essentially temporal and dynamic. That is the main thesis of this book.

Mental Features: The World-Brain Relation

Why care about the debates of time? These are all theoretical-philosophical debates. Why shall I care in my daily life about these issues about time anyways? This book tells you why! Time is the basis of your mind and mental life. There would be no consciousness, let alone a sense of self, without time. Being well-informed about the latest scientific advances, you may want to say that mental features like self and consciousness are based on the brain. Neuroscience is the hot trend as it reveals some of the remaining mysteries of our time, our mind. Neuroscientists tell us that mind is brain and that we no longer need to assume a mind as the old philosophers did.

The brain is all we need to explain the mind. However, there is a gap. How can the brain and its neuronal activity bring forth something like the mind that is seemingly distinct from the brain itself? We do not experience any neuronal activity in our consciousness. Also, we sense our self in a mental way rather than in neuronal terms: I do not experience my self as my brain. If mental features are based on the brain, we need to show how its neuronal activity transforms into mental activity. That process of transformation, which I term "neuro-mental transformation," remains elusive so far.

Let's compare the situation to water. Canada has harsh seasons. Having temperatures of about −40° Celsius in winter means that water is frozen, while it becomes fluid in spring and may evaporate in the hot summer. How can one and the same chemical transform into such different states? The temperature of the context, the environment, is key. Analogously so in the case of the brain: if the brain is well connected to its context, namely body and world, its neuronal states transform into mental states. Rather than conceiving the brain alone by itself, we need to take into view its relationship with the world, the "world-brain relation" (Northoff 2016, 2018). Like the weather with its various seasons provides the context for the state of the water in Canada, the world constitutes the context for the brain to yield mental features.

Mental Features Are Dynamic Features

The reference to world-brain relation again begs the question: What features in the world-brain relation make it possible to yield mental features? The answer points back to time. World and brain can be related to each other through their dynamic – the world-brain relation is essentially temporal and thereby provides the ground for transforming neuronal into mental activity.

Mental features are temporal features. Going back to the ancient Greek philosopher Heraclit, time changes all the time, it is dynamic. We never step into the same river twice. Analogous to the river, our brain's neural activity continuously changes so that we never encounter one and the same brain twice. And, as we all know only too well, the same holds for our mental features like self and consciousness. We never experience one and the same thing twice in our consciousness – there is continuous change in our experience. The same holds for our self which too continuously changes – we are never exactly the same person over time.

Change over time and thus dynamic is the core of our mind. If there is no change, the mind, including consciousness, disappears. The mind is intrinsically dynamic. The dynamic of the world-brain relation is key for constituting our mind. If that is lost, we lose consciousness. This is, for instance, the case in conditions like in coma or when under anesthesia. In short, mental features are dynamic features.

The Structure of This Volume

The core question driving this book is: How does the dynamic of time shape the connection of world and brain in such way that mental features like self and consciousness are yielded? Time is dynamic. The dynamic of the waves of time characterizes the world in general, just as it characterizes the ocean as part of the world. The same holds for the brain's inner time that is part of the world's outer time with its much larger range. The waves of the world pass through the brain. Such passing of the world's waves through the brain is central for yielding mental features like self and consciousness – the dynamic of world-brain relation is the very basis of our mind. That is the main message of this book.

In chapter 1, I describe the waves of the brain, and how they are constructed by the brain's neural activity. In chapters 2 and 3, I demonstrate how the brain's waves are synchronized with the ones of both body (chapter 3) and world (chapter 2) – the brain's waves align and integrate themselves with the time of body and world. There is continuous change in our consciousness – it is like a river or stream, the "stream of consciousness" as the famous American psychologist William James once said. Consciousness is based on the brain's role as neuronal surfboard that rides and surfs on the waves of both body and world.

Any surfer only knows too well, the surfboard does not surf by itself. One needs a surfer who directs and guides the board: the self, as I discuss in chapter 4. The surfboard continuously changes position. The surfer, in contrast, remains continuous and shows duration, that is, continuity over elapsed moments in time. I demonstrate how the brain's waves construct their own duration as basis for our sense of self. Yet another crucial feature of waves is their speed, as they can be faster or slower. Speed, in turn, is based on velocity and duration. In chapter 5, I show how the brain's construction of its speed and velocity can yield states like depression and mania featured by perception of extreme inner time speed, e.g., too slow or too fast.

Time beyond Humans – From Other Species over Artificial Intelligence to the Mind-Body Problem

Do other species like animals and artificial devices exhibit mental features? Animals can mentally surf on the waves of the world. Although they may use the world's waves in a slightly different way than us, as their brain's inner time is different, they use slightly different frequencies and speeds – they must thus show consciousness and sense of self which, based on their temporal overlap, partially overlaps with our human mind.

How about artificial intelligence (AI)? Despite all progress, AI devices do not yet exhibit an inner time which makes it possible for them to align and integrate the ongoing waves of the world's outer time – they lack the world-brain relation or better world-agent relation of humans. Therefore, I conclude in chapter 6 that AI devices remain currently unable to surf on the ongoing waves of the world's time; they therefore lack mental features like consciousness and self.

You now may want to ask about the existence of the mind as such. Philosophers have long argued whether a mind exists and, if it does, how it is related to the brain and the body. This amounts to what philosophers describe as mind-body problem. My answer to that question is simple: all we need to explain mental features is the world-brain relation and its temporal characterization in terms of waves featured by their dynamic.

The waves of the world-brain relation provide the ground upon which the brain can construct its own inner dynamic, the neuronal surfboard for our mind as the surfer: the better the brain as neuronal surfboard is adapted and aligned to the dynamic of the world, the better it can yield mental features like consciousness and self and thus navigate us in the world's often stormy waves. What is described as mind-body problem can then be traced to the dynamic relation of world and brain, the "world-brain problem" (Northoff 2016, 2018), as I conclude in chapter 6.

The Self as Wave

Surfers adapt to the waves of the sea through their surfboard. The more dynamic the surfer is the better they can adapt and align to the waves of the sea and the better the latter will propel the surfer forward and the longer the surfer will remain on their board. The same holds, analogously, in the case of the self. The more dynamic it is, the better it can adapt and align itself to its dynamic changing environmental context, and the more stable and temporally continuous it is. Through the brain's waves, the self is a wave by itself that thereby rides on the much larger wave of its environment.

1
Brain Time

Introduction

How are brain and time connected? This may seem like a strange question. Time is usually considered in physics and philosophy, while the brain is the focus of neuroscience. The discussion of time – what it is and how it is important for the brain – is thus not a major topic in neuroscience. But it should be at its very core. As for much of the world beyond neuroscience, time is key to understanding the brain and especially its connection to the mind. We therefore can take important lessons from the physics and philosophy of time to deepen that understanding.

The Container versus Construction View of Time

Time may be viewed as a container or as a construction. The "container view" conceives of time in terms of simple points at which certain events or objects occur (within the container provided by time). As these points are not part of and thus are extrinsic to the events or objects themselves, they are "outer time." This is the view of time that classical physics implies and current neuroscience considers dominant: the brain featured by its inner neuronal and ultimately mental events is "located" within the outer time of the world as its container.

In contrast, the "construction view" conceives of time in terms of relations between events or objects; these spatiotemporal relations are part of the events or objects themselves. As such relations are intrinsic (rather than extrinsic) to the events or objects themselves, they are "inner time." Such inner time is most apparent in quantum theory that describes continuous construction of time in terms of waves. The brain, too, exhibits an inner time, with its own waves that feature continuous change and passage. Such a construction of its own inner time distinguishes the brain from the outer time of the world – this entails a construction rather than container view of time. To show how the brain constructs the waves of its inner time, I turn to neuroscience.

The Brain's Inner Time

The Brain's Inner Time: Frequencies and Temporal Windows

The brain constructs its own inner time in terms of different waves, which can be short or long and thus show different "temporal duration." The temporal duration is related with the temporal ranges or cycle durations of neural oscillations or fluctuations. This includes different frequencies that range from infraslow (0.0001 to 0.1 Hz), over slow (0.1 to 1 Hz), delta (1 to 4 Hz), and theta (5 to 8 Hz), to faster frequencies like alpha (8 to 12 Hz), beta (13 to 30 Hz), and broadband gamma (30 to 240 Hz). These different frequencies show different functions and, most likely, are associated with different underlying neurophysiological mechanisms as well as distinct behaviours and functions.

The temporal duration of the brain's neural activity can be also characterized by intrinsic temporal autocorrelation in the milliseconds to seconds and minutes range. This means that the different sequential parts of a wave are related to each other over shorter and longer time distances. These timescales can be measured by an "autocorrelation window" and scale-free or fractal properties like the power law exponent or Hurst exponent (see below). This makes clear that the brain's "intrinsic" time (its "inner duration") is highly structured and organized. Such temporal structure and organization in the brain's neural activity is central for consciousness.

The intrinsic temporal organization of the brain's neural activity strongly influences the processing of extrinsic stimuli and events from the outer environment. The different frequencies with their respective cycle durations provide "windows of opportunity," or "temporal receptive windows" (Hasson et al. 2015; Wolff et al. 2022), to acquire and encode extrinsic stimuli/events through their temporal features. The brain constructs its own temporal receptive windows to receive and process the extrinsic stimuli/events from the outer world (Wolff et al. 2022; Golesorkhi et al. 2021a). Just like the different windows of a house allow an inhabitant to view distinct features in the surrounding environment, the brain uses its inner temporal windows to process the events in the outer environment.

The Brain's Inner Time: Scale-Free Activity

The brain's spontaneous activity shows a sophisticated temporal structure that operates across different frequencies from infraslow over slow and fast frequency ranges. Importantly, the neural activity in these frequencies shows a nested, or fractal, organization: just like a smaller Russian doll nests within the next larger one and so forth, the lower power of a faster frequency nests within the slightly higher power of the next slower frequency and so forth. Such temporal nesting is described as "scale-free dynamic," since the same temporal relation (of slow and fast frequency power) holds across different (i.e., slow and fast) timescales.

Scale-free dynamic is characterized by long-range temporal correlation (LRTC). LRTC describes that the slower frequency is related to and thus correlated with the faster frequency through their temporal relation. Since LRTC predominantly reflects irregular fluctuations in the infraslow frequency range (0.01 to 0.10 Hz), rather than the more regular oscillations in the faster ranges, it was often conceived as mere "noise." However, the noise-like signal derives from the neural activity that is organized in a specific way, that is, reflecting structured noise or "pink noise" (He et al. 2010). This can be distinguished from the noise related to our method of measurement, which is not structured and reflects "white noise" or "Poisson noise." The data in the next chapters suggest that the structured noise-like signal is central for both self and consciousness.

In addition to the scale-free fractal nature of LRTC, infraslow, slow, and faster frequencies also couple with each other in cross-frequency coupling (CFC). CFC refers to a systematic relationship across frequencies where the (phase of the) slower frequencies couple to the (amplitude of the) faster ones in an upward progression. That again provides a certain structure or organization among the different frequencies.

In sum, the brain's inner time can be characterized by scale-free activity which constructs an intrinsic or inner temporal duration in its neuronal activity. As I show below, the brain's construction of its own inner temporal duration is central for mental features like self and consciousness. Moreover, scale-free activity is ubiquitous in the world. I argue that the brain's temporal duration, constructed through its scale-free activity, is nested within the much more extended temporal duration of the world.

From the Brain's Inner Time to the World's Inner Time

How is the brain nested within the world? As in the brain, activity in the world fluctuates over time. These fluctuations occur in different frequency ranges, for instance from 0.01 Hz to 240 Hz (or otherwise), as we can observe them in the brain (as part of the world). Importantly, these fluctuations are not completely random, but show a certain relationship or structure with each other across time. This temporal structure operates across different timescales or frequencies, hence the name scale-free activity. It can be described by temporal nestedness: the more powerful slower fluctuations contain the less powerful faster frequencies, just like a larger Russian doll contains the next smaller one.

One example of temporal nestedness in the world are seismic earth waves. The Chinese American neuroscientist Biyu He (He et al. 2010) investigated not only the scale-free dynamics of the brain, but also the time series of activities from spontaneous earth seismic waves collected over four months and fluctuations in the Dow-Jones index collected over eighty years. Her question was whether the time series in earth seismic waves and stock market fluctuations also show scale-free activity with LRTC indexing temporal nestedness.

Just as in the case of the brain, time series from both earth seismic waves and stock market fluctuations followed scale-free dynamic in their temporal power spectrum. Interestingly, their PLE (1.99 for seismic waves and 1.95 for the stock market) came close to that of the brain's spontaneous activity (mean of 2.2 for < 0.1 Hz). Given these and other examples in nature, like sea waves, wind, and birdsongs, scale-free activity is a universal feature in the world. The world constructs its own inner time in a scale-free way with temporal nestedness between the fluctuations of its different frequencies. The same holds true for the brain. As the brain is part of the world as whole and its scale-free activity, the brain also constructs its own inner time in a scale-free way.

The scale-free construction of its own inner time makes the brain part of the world as whole and integrates it within the world's inner time: the brain's inner time is nested within the world's inner time in a scale-free way. Temporal nestedness of the brain within the world implies that there are LRTC between the world's inner time and the brain's inner time. The brain's LRTC thus lock us within the world – such a state is key for consciousness. Breakdown of the brain's scale-free dynamic with its LRTC detaches us from the world – we are then locked out of the world, which leads to the loss of consciousness (see chapter 2).

Convergence of Time and Space

The Brain's Inner Space: Spatial Structure of the Brain's Spontaneous Activity

This book focuses on brain and time. However, physics and philosophy tell us that time is intimately connected to space. The same is true in the case of the brain. Therefore, I will briefly discuss the brain's spatial structure and its "inner space" as featured by "spatial extension" and how it relates to temporal duration.

Early neuroimaging investigation using techniques such as fMRI and EEG focused on stimulus-induced activity characterized by the brain's response to sensorimotor, cognitive, affective, or social stimuli or tasks. Recently, neuroimaging has shifted focus to the brain's spontaneous activity featured by its spatial and temporal structure. Initially, it was thought that spontaneous activity was contained to a particular neural network, the default-mode network (DMN). However, it soon became clear that spontaneous activity is pervasive throughout the whole brain.

Spontaneous activity can be observed in many different neural networks, including what is described as the central executive network, the salience network, and the sensorimotor network. Even in regions as dependent on external stimuli as the sensory cortices, there is spontaneous activity. Specific regions coordinate their ongoing resting-state activity levels with each other – this is measured by their functional connectivity, thereby forming neural networks. Together, the relationship among regions resulting in networks constitutes an elaborate spatial structure and extension within the brain's spontaneous activity.

The Brain's Inner Space: Convergence of Spatial Extension with Temporal Duration

How are spatial structure and extension related to the temporal structure and duration of the brain's intrinsic activity? In one study, de Pasquale et al. (2012) observe that the DMN (and especially the posterior cingulate cortex) shows the highest degree of correlation with other networks in specifically the beta frequency range. The DMN seems to interact much more with the other networks than the latter do with each other. The reasons for that remain unclear but may, in part, be due to the central position of the DMN (and its midline structures) in the middle of the brain.

Such cross-network interaction is dynamic and transient, and therefore continuously changing. There are, for instance, alternate periods of low and high synchronization between the DMN and other networks. These findings suggest that the spatial structure is closely linked to temporal dynamics, that is, oscillations in different frequency ranges. Specifically, different neural networks may show different frequency ranges. For instance, Hipp, Hawellek, Corbetta, Siegel, and Engel (2012) observe that the medial temporal lobe is mainly characterized by theta frequency range (4 to 6 Hz), the lateral parietal regions are characterized by alpha to beta frequency range (8 to 23 Hz), and the sensorimotor areas show even higher frequencies (32 to 45 Hz). These findings demonstrate the close link between spatial and temporal dimensions in the spontaneous activity.

The close link of spatial and temporal structure is also manifest in the temporal windows. Shorter temporal windows in the brain's neural activity are mainly located in sensory and motor regions. This is highly suitable as these regions need to process the fast inputs and mediate the fast motor outputs – for that, shorter temporal windows allowing for high temporal precision are optimal (Wolff et al. 2022; Golesorkhi et al. 2021a; Golesorkhi et al. 2021b). These temporal windows keep us on track when we need to perform external tasks, like driving a car during rush hour.

In contrast, the regions in the core of the brain like the default-mode network exhibit longer temporal windows: they are not as involved in the processing of external inputs, but more strongly mediate internal inputs. These windows keep us on track when we dwell in our own internal thoughts rather than paying attention to external inputs, like driving a car on long empty stretches of highway.

The longer time windows are thus mainly in the core or middle networks of the brain as they are not as exposed as strongly to the external inputs like the sensory regions and networks. Thus, temporal duration and spatial extension converge in the brain – this is key for understanding mental features like self and consciousness.

The Active Brain and Its Inner Time

A Passive Model of the Brain: Perception and Cognition of Outer Time

The brain actively constructs its own inner time. That is very different from the view of the brain in past and present neuroscience. There, the brain is conceived more like a passive organ that receives and processes the stimuli and events of the outer world. On the philosophical side, such passive view of mind was proposed by the British philosopher David Hume (1711–1776). He conceives the mind with self and consciousness to result from mere association of external stimuli or events. In contrast, the mind itself does not show any intrinsic or spontaneous activity apart from the one caused by the extrinsic inputs from the environment.

Such passive view of mind has been transposed to the brain. The early British neurologist Sir Charles Sherrington (1857–1952) proposes that the brain and the spinal cord are primarily reflexive. "Reflexive" in this model means that the brain is merely passive as it reacts in predefined and automatic ways to sensory stimuli such as auditory or visual stimuli. Stimuli from outside the brain, originating externally in either the body or the environment, are assumed to determine completely and exclusively the subsequent neural activity. The resulting stimulus-induced activity, and more generally any neural activity in the brain, is traced back to the external stimuli and events themselves. The brain is thus completely determined by the outer time of the environment. In contrast, it shows neither intrinsic activity nor inner time – I therefore speak of a "passive model" of brain.

The passive model of brain is also prevalent in our time. Neuroscience focuses on the neural correlates of our perception and cognition of events in the world. Time is here associated only with the external stimuli and events themselves, that is, their discrete moments or moments in time: the brain simply processes the stimuli's discrete moments in time and, at best, associates and integrates them (Wolff et al. 2022). Time remains purely extrinsic to the brain – the brain does not possess its own inner time as distinct from the world's outer time, which then provides the container for the former. In contrast, there is no focus on the brain's construction of its own inner time in current neuroscience.

An Active Model of Brain: Spontaneous Activity

Such a passive view of the brain conflicts with our experience of the active nature of our mind, though. We experience free will, as well as spontaneous thoughts and dreams, independent of any external stimuli or events. How can we account for the active nature of our mind given the passive view of the brain? To connect the brain and the mind, we need to link the seemingly passive brain with our experience of an active mind. One strategy is to view the brain's active construction of time and how that is key for mental features.

Hume was followed by the German philosopher Immanuel Kant (1724–1804). Kant suggests that the mind is not merely passive by processing and associating external stimuli. Instead, the mind adds something, that is, its own inner activity. The mind possesses its own spontaneous activity featured by its inner time and space – Kant characterizes such spatiotemporal structure by what he describes as the "categories." Importantly, the mind's own inner spatiotemporal structure shapes how it processes external stimuli – the activity induced by external stimuli and events is thus a hybrid of both active and passive contributions.

The Kantian active model of mind suggests that the brain itself actively constructs its own spontaneous activity featured by inner time independent of external stimuli (Northoff 2012). Such a model had already been suggested by one of Sherrington's students, Thomas Graham Brown (1882–1965). In contrast to his teacher, Brown suggests that the brain's neural activity – that is, within the spinal cord and brain stem – is not primarily driven and sustained by external stimuli. Instead, Brown holds that neurons in both the spinal cord and the brain stem show spontaneous activity that originates internally, something which nowadays is known as central pattern generators in motor neurons. Besides proposing a more active view of the brain, he was very active in alpine climbing and detected three novel routes up the east face of the Mont Blanc.

Other neuroscientists followed Brown in the first half of the twentieth century. Hans Berger (1873–1941), a German psychiatrist best known for introducing electroencephalogram (EEG) in 1924, also observed spontaneous activity in the brain. Unfortunately, he could not see the fruitfulness of his own discovery as he committed suicide after suffering from depression during the Second World War. Other neuroscientists like George H. Bishop (1933), Karl Lashley (1951), and Kurt Goldstein (2000) follow in Brown's line of thought and propose that the brain actively generates its own spontaneous activity. While the brain's spontaneous activity was largely neglected in the second half of the twentieth century, it experiences a resurgence with the discovery of the DMN by Marcus Raichle

(Raichle et al. 2001), one of the leading neuroscientists of our time. Raichle demonstrates high levels of spontaneous activity in specifically the DMN, which he supposes to serve as default-mode or baseline for the brain and its cognition (Raichle 2009; Raichle 2015; Northoff et al. 2022).

From Brain to Mind: Inner Time as Their Common Currency

Together, these and other observations point to the key role of the brain's spontaneous activity as it actively shapes how external stimuli and events are processed. This leads me to speak of an "active model" of the brain. The early German neurologist Kurt Goldstein in his 1934 book *The Organism* illustrates such an active model:

> The system is never at rest, but in a continual state of excitation. The nervous system has often been considered as an organ at rest, in which excitation arises only as a response to stimuli. ... It was not recognized that events that follow a definite stimulus are only an expression of a change of excitation in the nervous system, that they represent only a special pattern of the excitation process. This assumption of a system at rest was especially favored by the fact that only the external stimuli were considered. Too little attention was given to the fact that the organism is continuously exposed, even in the apparent absence of outward stimuli, to the influence of internal stimuli – influences that may be of highest importance for its activity, for example, the effect of stimuli issuing from the blood, the importance of which was particularly pointed out by Thomas Graham Brown. (Goldstein 2000, 95–6)

The observation of the brain's spontaneous activity profoundly shifts our view of it. Rather than conceiving of the brain as a purely extrinsically driven device, the spontaneous activity suggests what Raichle describes as an "intrinsic model of brain" (Raichle 2009; 2010). This is reminiscent of a Kantian-like model: the brain's spontaneous activity structures and organizes its own task-evoked or stimulus-induced activity, including related sensory and cognitive functions (Northoff 2012a; 2012b). That converges with the active model of brain.

The active model of brain shifts the focus to how the brain constructs its own inner time and space. I showed, for instance, the brain's construction of temporal duration through autocorrelation and scale-free dynamic. What is the function and role of the brain's inner time? We currently do not know. I aim to demonstrate that the brain's inner time is key in constituting mental features. Specifically, I propose that the spontaneous activity's construction of its own inner time (and space) provides a hitherto missing link of brain and mind: both neural and mental states share the brain's inner time as "operational time" (Fingelkurts et al. 2010) or as their "common currency" (Northoff et al. 2020a; Northoff et al. 2020b), as I demonstrate in subsequent chapters.

Conclusion

Brain and time are usually dealt with separately in neuroscience and physics. However, I here converge both as I characterize the brain by its own inner time as distinct from the world's outer time. This is empirically strongly supported. Recent studies show that the brain actively constructs its own inner temporal duration through autocorrelation and scale-free dynamic. Following historical predecessors, this leads me to speak of an active model of the brain as distinct from the predominating passive model of the brain.

The active model characterizes the brain by its own inner time and thus by continuous change, or dynamic. Paraphrasing the ancient Greek philosopher Heraclit (535–475BC), "we never encounter twice one and the same brain." This entails what is described as dynamic view of time in the current philosophical debates; that is, events change position in relation to the present moment like the sunset moves from the future to present to past. Such a dynamic view is distinguished from a static view in which time does not change at all.

Following recent views of time in physics (Smolin 2013; Weinert 2013; Rovelli 2018), I here suppose a dynamic view of time of brain and world. In the next chapter, I will show how the dynamic of the brain's inner time shapes its relationship to both the body and the world. That, in turn, is key for constituting mental features like self and consciousness; brain and mind share the same dynamic as their common currency (Northoff et al. 2020a; Northoff et al. 2020b).

2
From Brain Time to World Time

Introduction

Time can be constructed in different ways in the world. I have shown that time is constructed in seismic earth waves over extreme long periods of ultraslow fluctuations until they suddenly erupt into an earthquake. Modern physics aims to understand and investigate the various constructions of inner time by seismic earth waves. This is possible by, for instance, investigating the time series beneath the seismic earth waves where, following its changes over time, specific principles can be extracted. One such principle is scale-free dynamic as featured by temporal nestedness of different frequency ranges (slow-fast) and their respective timescales (long-short). Importantly, this is true for the construction of time in both the world and the brain.

How about consciousness? On the one hand, if the brain's inner time is relevant for consciousness, one would expect scale-free dynamic, as one of its key principles, to be altered if not lost in those states when consciousness is reduced or even lost. That is, for instance, the case in anesthesia, sleep, or coma. On the other hand, there are also states where our consciousness is increased and we are hyperaware, such as when taking psychostimulants like LSD, psilocybin, ketamine, or ayahuasca. Neuroscientists have investigated these states using brain imaging techniques like fMRI or EEG, as I describe below.

In this chapter, I will demonstrate the key relevance of the brain's inner time for consciousness. Connection of the brain's inner time to the world's inner time is necessary for consciousness. If, for instance, we are locked out of the world and its time, we lose consciousness. The connection of the brain to the world, their dynamic tango, is thus indispensable for consciousness – even though current approaches to consciousness tend to focus on the brain itself or, as in the embodiment approach, on brain and body.

Brain Time: Scale-Free Dynamic and Consciousness

Breakdown of Scale-Free Dynamic

Scale-free dynamic describes the balance of power, with stronger power in slower frequencies relative to faster ones (see chapter 1). This creates long-range temporal correlation (LRTC) across the different frequencies and their respective timescales. The LRTC can be measured by power law exponent (PLE) or detrended fluctuation analysis (DFA).

How is scale-free activity related to consciousness? It refers to the degree of arousal and wakefulness of a subject. During anesthesia, sleep, or coma, the level of consciousness is decreased. In contrast, while taking psychostimulants or during sleep deprivation, the level of consciousness is increased and subjects show greater vigilance and sensitivity to external perturbations. For instance, when one is very tired but cannot sleep, one is hyperexcited. One's processing is very fast and the perception of external events or objects may be extremely sensitive. That, in turn, may lead to high levels of excitement with strong reactivity and distraction.

How about scale-free activity, as indexed by PLE or DFA, in these various conditions? Several resting state studies (using mostly fMRI) demonstrate extremely low if not completely disrupted indices of scale-free activity in subjects during deep surgical anesthesia, deep sleep, and coma (Zhang et al. 2018; Tagliazucchi et al. 2013, 2016). In that case, the power in both slow and faster frequencies is extremely reduced, which basically leads to a complete breakdown of the scale-free organization including the LRTC in the brain's spontaneous activity. The power spectrum is flat, with equal power distribution (low power) among slower and faster frequencies. This means that there is no difference anymore between slower and faster frequencies or their respective longer and shorter timescales.

Breakdown of Consciousness

Why does the loss of frequency/timescale difference in the brain's inner time lead to the loss of consciousness? Any processing of inputs including both internal from body/brain and external from environment is now the same. The brain can no longer add any different frequencies or different timescales to the processing of the different inputs. Regardless of where the input comes from – brain, body, or environment – or what it contains, all inputs are processed in the same way. This compares to the situation when one continuously speaks the same word or sentence all over again and again. There is no differentiation anymore. The brain's neural activity is one homogeneous soup without any temporal (and spatial) differentiation, and the temporal structure of its inner time is lost completely.

If our brain's inner time shows no longer any temporal structure and differentiation, any kind of differential experience of the world, including the difference of self, body, and environment, is also lost. With no differentiation, we experience only a homogeneous soup, which ultimately means we do not experience anything at all. Consciousness is lost when the brain loses the scale-free dynamic structure of its inner time – we are locked out of it.

How about cases where consciousness is clouded, such as in sedation, earlier and lighter sleep stages (like N1 and N2), and minimally conscious state (MCS)? In those cases, the power in slow frequencies is preserved or increased, while the power in faster frequencies is decreased or diminished. There is consequently increased PLE or DFA. Scale-free activity in these cases of clouded consciousness is thus preserved but abnormally shifted towards the slower frequencies. The brain in these states can thus be compared to a person who is walking or swimming but doing so on autopilot. By comparison, in cases where consciousness is absent, a person can no longer walk or swim at all – any movement is completely broken down, as manifested in a corresponding total breakdown of consciousness.

Extension of Consciousness

How about changes in the spontaneous activity's PLE or DFA in those conditions that exhibit an increased level/state of consciousness? Recent studies (mostly using fMRI) on drug-induced psychosis using psychostimulants demonstrate increased power in the faster frequencies, while the power in the slower frequencies remains the same. The relative power balance thus shifts toward the faster frequencies, which results in lower PLE or DFA values in drug-induced psychosis. More or less analogous results with increased faster frequency power and lower PLE or DFA were also observed in EEG during sleep deprivation (Meisel et al. 2017).

Due to such constellation with a shift of scale-free activity towards faster frequencies, the brain's dynamic repertoire is increased, rather than decreased as when consciousness is lost. Opportunities are vastly extended now for the brain's dynamic repertoire, but that also creates problems such as how to control such extended repertoire. That extension of the brain's dynamic or temporal repertoire manifests in the extended spatial and temporal boundaries that subjects experience during such states of extended consciousness. Their subjectively perceived spatiotemporal repertoire is now extended, to which some subjects react with comfort and others with anxiety.

Taken together, the data show opposite changes in the spontaneous activity's PLE or DFA during increases or decreases in the level/state of consciousness. Decreases in the level/state of consciousness, such as in sedation, N1 or N2 sleep, and MCS, lead to increased PLE or DFA as the power spectrum shifts from faster to slower frequencies. The opposite can be observed in increases of the level/state of consciousness; here the power spectrum shifts towards the faster frequencies resulting in decreased PLE or DFA.

As discussed, these changes lead to opposite changes in whether consciousness is clouded or extended. We can now start to see how the brain's dynamic or temporal repertoire with its scale-free activity directly translate into corresponding temporal changes in consciousness – time, or scale-free activity, provides the common currency of brain and consciousness (Northoff et al. 2020a; Northoff et al. 2020b). More generally, these examples make it clear that time is the very basis of consciousness.

Time includes both continuous change and presence. The continuous change and flow of the brain's inner time across its different frequencies and timescales first and foremost make possible the presence of consciousness. Change and presence are thus not mutually incompatible, but compatible: our brain's neural activity is like a river that is subject to continuous change in a certain structured way along scale-free dynamic which, in turn, makes possible the continuous presence of consciousness. Once the river and its flow of water stops, the river and thus consciousness break down. Time and its structure are of essence not only for the river but also for consciousness.

Brain + World = Consciousness

"Self-Similarity" of the World and the Brain

Why is scale-free activity relevant for consciousness? For that we need to dive one more time into the nature of scale-free dynamic. The fluctuations across the different timescales are not merely random or coincidental. Instead, they follow scale-free dynamic distribution. Scale-free dynamics means that no specific scale dominates the dynamics of the activity or process. This is manifested in correlations across LRTC that decay more slowly and extend over longer distances in time. There is a certain temporal structure that signifies self-affinity or self-similarity.

"Self-similarity" describes when a part of a whole is an exact fractal of the whole. Purely mathematical and geometrical fractals can be characterized by self-similarity. Another instance is Russian dolls nesting within each other; they are self-similar, with each doll an exact replica of the original, and while the largest may be on a different scale in spatial regard thus is scale-free, it is still nevertheless self-similar.

For another example of a self-similar system, consider the Roman cauliflower (see Hardstone et al. 2012). Exact copies of the entire cauliflower are recognizable on multiple smaller scales. There is no particular size of cauliflower floret that can be characterized by scale-free dynamics. Most importantly, there is an inverse relationship between the frequency of a certain size of floret and the size itself: the smaller the size, the more often it can occur. There are many more smaller scales of the cauliflower than its larger sizes. This entails scale-free activity according to the power law exponent (see chapter 1).

Scale-free activity is about self-similarity. The data show that the brain's scale-free activity modulates consciousness from total breakdown over sedation to "normal" and extended consciousness. Given these findings, I claim that consciousness is about self-similarity between the brain and the world: the more the brain's scale-free activity as an index of the brain's inner time is nested within and thus self-similar to the world and its inner time, the higher the level or state of consciousness. Put in a nutshell, consciousness is based on self-similarity and temporal nestedness of the brain and the world.

The brain's scale-free activity nests and thereby locks us within the world. Such virtual scale-free nesting positioning of the brain within the world makes it possible that consciousness can provide information and knowledge about the world. We experience ourselves and our body as part of the wider world as whole – which is only possible if our brain positions, and thus locks, us within the world.

However, sometimes we make errors in our judgment about the world; in that case, our brain does not lock us within the world in an optimal degree, that is, in a completely self-similar way, as in the case of the Russian dolls. Instead, the brain locks us to the world in an imperfect way, a self-affine way, as it is described. Consciousness illustrates that imperfect or self-affine match between the world and the brain. The better and more perfectly the brain locks and matches us with the world, the more accurate our consciousness. Experienced meditators know this only too well: better alignment with the world and concurrent detachment from one's own perceptions, cognitions, and body only increase consciousness – this is experienced along with an increased flow of time and alignment with the environment, unperturbed by any personal contents.

The Correlated Walk of the World and the Brain

How do we have to imagine such alignment of the brain's inner time to the world's inner time? Hardstone et al. (2012) introduce the example of different kinds of walkers. A random walker will arbitrarily choose which direction, right or left, to take at each intersection on their path, and so will go sometimes left and sometimes right (the Hurst exponent will be 0.5; see below). An anti-correlated walker, in contrast, will choose their direction based on which is opposite to their previous direction: left will follow right, right will follow left, and so on (the Hurst exponent will be < 0.5). Finally, a correlated walker will prefer either direction, left or right, throughout their walk: they will preferably take the left, but not always (the Hurst exponent will be > 0.5).

Compare the direction the walker takes at each intersection to the spatiotemporal activity patterns in different systems of the natural world, including the brain. Very much like our walker, the activity patterns can take different directions at specific points in time and space. Showing self-affinity (e.g., self-similarity), they can be compared to the correlated walker, the one who prefers a particular direction over time. This leads to LRTC in the walking pattern across time.

Why are these examples relevant? Exhibiting scale-free activity, the relation between the world's inner time and the brain's inner time can be compared to the "correlated walker in space and time." World and brain walk with each other in a more or less correlated way in space and time – there is a certain degree of temporal correspondence of their respective inner time constructions, i.e., their scale-freeness. The more the world's and the brain's inner time correspond and correlate, the stronger they correlate with each other over time, and the more they share LRTC as form of "dynamic structural memory," as Linkenkaer-Hansen et al. (2001, 1375 and 1376) demonstrate in their seminal paper.

As it is dependent upon the brain's scale-free activity including its relation to the scale-free activity of the world, consciousness reflects our brain's "correlated walk in the space and time of the world." In the same way that the correlated walk helps the walker to navigate their path, consciousness helps us to navigate the world: consciousness allows us to find and navigate the way within the world that is most appropriate to us and our individual brain. That is because the scale-free activity allows for a correlated walk of the brain and the world with shared dynamic structural memories.

Without consciousness, we are lost in the world. In that case, there is no longer any correlated walk of the brain within the space and time of the world. The brain is locked out of the world as is manifested in the complete absence of consciousness. We are then no longer part of the world as whole and therefore cannot navigate ourselves within it. The opposite case occurs in extended consciousness. Psychostimulants induce too many options for a correlated walk when extending our consciousness; we are too attached to and locked in the world. Some people can handle and enjoy the experience of such high correlation with the world. Others cannot handle such vast increase in the dynamic repertoire or options for walking and contents – this can result, in the most extreme instances, in what is described as a "horror trip."

Mental Surfing on the World's Waves

How can we imagine such correlated walk of brain and world? Let's compare that to surfing. Consciousness is our way of surfing on the world's waves as they pass through our brain. The better the surfer aligns their surfboard to the ongoing scale-free fluctuations of the ocean waves, the better and longer they can surf and stand on it. The same applies analogously to the world and our consciousness. The better our brain aligns itself to the world's waves, the better and the longer we can surf on those waves and maintain our consciousness.

If our consciousness becomes clouded and thus more restricted, the surfboard shrinks until it completely breaks down such that the surfer cannot but fall into the water; that is the moment we become completely detached from the world as manifests in the loss of consciousness. If, in contrast, our consciousness extends, as when taking psychostimulants, the surfboard becomes extended such that our interface with the world is more extended – one consequently feels and experiences themselves as part of the wider and broader wave and ultimately the ocean itself with its dynamic.

Accordingly, consciousness is the surfboard on which we surf on the world's waves. For that to be possible, we need our brain and its waves to align well to the ongoing change of the world's waves – the world-brain relation and its dynamic make consciousness possible. Therefore, unlike most other theories of consciousness, the temporo-spatial theory of consciousness (TTC) highlights the temporo-spatial alignment of the brain to the world as one key mechanism of consciousness (Northoff and Zilio 2022a; Northoff and Huang 2017; Northoff and Lamme 2020).

"Locked in the World" versus "Locked out of the World"

From World to Brain: Consciousness Extends beyond Brain and Body

Consciousness extends beyond brain and body. We are conscious of the world and experience ourselves, including our body, as part of it. Rather than being restricted to the brain or the body, consciousness extends beyond ourselves to the world. Our point of view is "located" within the world and that, in turn, makes possible the subjective nature of consciousness as manifest in its first-person perspective (Northoff and Smith 2022). How is such extension of our point of view as basis of consciousness beyond ourselves possible?

One way is that the brain aligns with the world in more or less the same way that it aligns to the body: in temporal terms. Such temporal alignment of the brain to the world would, in turn, make it possible to extend consciousness beyond ourselves to the world. I will now show that this is exactly the case: the loss of the brain's temporal alignment with the world leads to the loss of consciousness. The brain processes external stimuli from the world in its sensory functions. These sensory functions align the brain to the world in a temporal way. For instance, when we listen to music, we may spontaneously move according to its rhythm; this is only possible if our brain's sensory and motor functions align their temporal structure with the temporal structure of the music.

The regions that are central to mediating sensory functions are the cortex of the brain and its connection to a set of subcortical nuclei beneath it, namely the thalamus. These thalamo-cortical connections are central for relaying and processing sensory stimuli from the external world within the brain. Importantly, as long as the thalamo-cortical connections are preserved, our consciousness of the world is lively and present. Even if the motor functions are lost, as long as the sensory functions with their thalamo-cortical connections are preserved, consciousness is maintained – as is the case in patients who experience locked-in syndrome (LIS).

Locked in the Body but Not Locked out of the World

What is LIS? The first scientific mention of a clinical situation resembling LIS was described in 1941 and subsequently diagnosed as akinetic mutism (Cairns et al. 1941). LIS is a situation in which the body is paralyzed without any motor action possible, while consciousness and cognition are fully intact. Though their body is impaired, the patient maintains basic cognitive abilities, awareness, sleep-wake cycle, and capability of meaningful, though minimal, behaviour.

The example of LIS seems to suggest clear separation between the brain and the body: the brain is still working while the body is depleted. Such brain-body dualism suggests that the brain and its neuronal activity are by themselves sufficient for mental features like consciousness. Following the example of LIS, one might therefore assume that consciousness is in the brain or is identical to some of its neuronal patterns. But this is not the case. Indeed, the motor functions are depleted and remain absent in LIS.

This, importantly, does not entail the loss of consciousness. The patient's sensory functions are still preserved, and they can thus still align themselves to the world, i.e., temporal alignment. Several investigations show that sensory processing in especially thalamo-cortical connection and sensory cortices is well preserved in these patients – that is why their consciousness is maintained. Accordingly, though locked in their body, LIS patients are not locked out of the world (I borrow these expressions from Federico Zilio and a common paper of ours; Northoff and Zilio 2022a; 2022b).

Consciousness is not "in the head." It is not purely neuronal, but neuro-ecological – it is based on the brain's relation to the world. This is possible through the brain's temporal alignment with the world and its temporal structure. I therefore speak of a dynamic world-brain relation. The case of LIS demonstrates that the dynamic world-brain relation is central for consciousness. Without the brain's dynamic relationship to the world, consciousness remains impossible.

Accordingly, the brain dances tango not only with the body, but also with the world. During the tango, the music and its waves find their way into the body of the dancers. The same applies to world and brain. The waves of the world enter and pass through the brain and its waves – the brainwaves are miniatures of the much larger worldwaves. If the brain and its waves become detached or decoupled from the worldwaves, we lose consciousness. Consciousness is thus nothing but the manifestation of the brainwaves to the worldwaves.

Coma and Vegetative State: Sensory versus Motor Functions

Are sensory functions with their neuro-ecological world-brain relation really necessary for consciousness? I so far have only showed that motor function with brain-world relation is not necessary for consciousness. This is supported by the observation that LIS patients who show deficient motor function but preserved sensory function do not lose consciousness.

What about the reverse scenario? In that case, sensory functions would be disrupted while motor function remains intact, indicating that sensory functions and thus temporal alignment are necessary for consciousness. This scenario can occur in, for example, instances of coma, vegetative state, or unresponsive wakefulness state (URWS), such as after a motorcycle accident or other incident in which someone sustains severe traumatic brain injuries. In that case, patients do not show any consciousness and thus no relation to the environment, only exhibiting reflex-like, unintentional, and involuntary reactions. Often, such patients show changes in their thalamus, including in its connections to the cortex. The sensory functions are severely impaired, while the motor function are preserved (see Zilico and Northoff 2022a; Zilico and Northoff 2022b). No consciousness is present.

The case of URWS can be considered as more or less the reverse of the case of LIS. URWS shows disruption in sensory functions while motor functions are preserved. LIS shows disruption in motor functions while sensory functions are preserved. The consequences for consciousness are vastly different, if not opposite, as it is preserved in LIS and depleted in UWS.

Locked out of the World versus Locked in the Body

What does this reveal about the conditions that are necessary to maintain consciousness? Sensory functions and their thalamo-cortical connections are necessary since their loss, as in URWS, leads to the loss of consciousness. Once we lose our brain's sensory functions, we are locked out of the world as we lose consciousness. Sensory function and the brain's temporal alignment with the world are thus necessary for consciousness. Therefore, consciousness is neither in the head nor in the world. Instead, consciousness is based on the world-brain relation – consciousness is relational and neuro-ecological.

In contrast to sensory function, motor function does not seem to be necessary for consciousness. We can be locked in our body, as in LIS, without being locked out of the world. The brain's relation to the body, the brain-body relation, and ultimately the brain-world relation, as presupposed in movements and motor function, is not a necessary condition of consciousness. Otherwise, LIS patients would lose consciousness. Even if we are locked in our body, our brain, based on its sensory functions and thalamo-cortical connections, can still align with the world through its own temporal dynamic. Accordingly, being locked in the body does not entail the same loss of consciousness as being locked out of the world.

Together, this amounts to an asymmetry between the sensory-based world-brain relation and the motor-based brain-world relation. The loss of sensory functions and the consecutive loss of world-brain relation leads to the loss of consciousness – being locked out of the world. In contrast, the isolated loss of motor functions with its brain-world relation locks one within the body, but does not lock one out of the world. Consciousness is thus intimately connected to our brain's temporal alignment with the world.

Conclusion

World is time and time is world. Since the world is time, the brain, as part of the world, is also time. Consciousness is time. Without time, there would be no consciousness. Consciousness has it own inner time. That inner time of consciousness is based on the inner time of the brain which, in turn, is related to and integrated with – nested within – the inner time of the world. Hence, consciousness is about the inner time as it is shared by world and brain. Accordingly, time (and space) provides the common currency of world, brain, and consciousness (Northoff et al. 2020a, 2020b).

We are part of and integrated within the world in our consciousness such as that we can navigate within the world and its inner time. If consciousness is lost, we are locked out of the world. That, as I have demonstrated in this chapter, is because the brain itself is then locked out of the world as it can no longer construct its own inner time in a scale-free way and thus in the way the world constructs its own inner time.

Consciousness is nothing more than the surfboard on which we ride the continuously changing waves of the world. The seemingly contradictory features of time, change, and presence are intimately linked in consciousness. The continuous presence of our mental surfing depends on the continuous change of the oceans' ongoing waves, the world waves. In the same way that a surfer experiences a groove if they can surf well and ride long in the ocean's wave, we experience a high or groove if we are well aligned to the world's waves. That mental groove is best illustrated by the extended consciousness induced by psychostimulants or meditation.

3
The Tango of Brain Time and Body Time

Introduction

The brain is not an isolated organ. It is closely interwoven with the body and the world. We feel our emotions within our bodies. We synchronize with music in the world by making rhythmic movements while, for instance, dancing tango. Does time mediate the relationship of our brain with the body and the world?

Like the brain, the body and the world each have their own inner time. In the body, there is the time of our heartbeat, our respiration, and even our stomach. In the world, consider the Pacific Ocean around Hawaii: there are many smaller, weak waves, some medium waves, and only a few big and extremely powerful waves. The surfer looks for the medium waves to provide a good ride, but the big waves are too powerful and ultimately too dangerous. The brain, like the surfer, prefers to ride on the medium waves. The big waves seem to be too big and dangerous for the brain – like the surfer, it is not properly equipped for them.

The surfer aligns their body to the dynamic of the waves. This is made possible by the brain that directs the body in the proper temporal and spatial way. I therefore speak of temporo-spatial alignment as one key mechanism of consciousness (Northoff and Huang 2017; Northoff and Zilio 2022a; Northoff and Zilio 2022b). Temporo-spatial alignment allows the brain to align itself with the environment through the body. I will now discuss several examples of such alignment and how it is key for consciousness.

Tango Time: Brain and Body

How is the brain aligned with the body? Imagine two tango dancers who align with each other, in that they adapt the rhythm of their movements and actions to each other. That is best possible by anticipating each other's movements, which is where the brain comes in. Through its own inner time, the brain conducts the body's movement and rhythms to synchronize and align with the rhythm of the music. The dance is thus a mixture of the brain's inner rhythms and the tango's outer rhythms. Consider a very special tango: the one between heart and brain.

Heart and Brain

Using functional magnetic resonance imaging (fMRI), various studies demonstrate that the dynamic changes in neuronal activity, i.e., functional connectivity between different regions (like from the amygdala and anterior cingulate cortex to subcortical regions like the brain stem, thalamus, and putamen, as well as to the dorsolateral prefrontal cortex), are directly related to changes in the heartbeat. The more variability in the heart rate, the more variability in the functional connectivity between these regions. Together, these fMRI studies demonstrate a close dynamic relationship between the brain's spontaneous activity and the heartbeat, that is, "neuro-cardiac coupling."

Lechinger et al. (2015) reported an EEG study on the relationship between heart rate and the brain's spontaneous activity during awake and asleep states. As I discussed in the previous chapter, the spontaneous activity of the brain continuously fluctuates; there are fast and slow fluctuations, as well as high and low activity periods in the fluctuations. These high and low periods form cycles of activity which are also described as phases. The beginning of one cycle marks the onset of a new phase featured by highs (peak) and lows (trough). Do the phase onsets of the brain's fluctuations correspond in their timing to the onset of the heartbeat?

If they do, one would assume that the brain's shifts its phase onsets to those of the heartbeat – this is called "phase shifting" or "phase locking." Lechinger et al. (2015) observe that the phase onset of especially the delta/theta frequency (2 to 6 Hz) in the brain's spontaneous activity locked to the onset of the heartbeat. Most interestingly, the phase locking of the delta/theta frequency to the heartbeat lessens progressively during the different sleep stages (N1 to N3) within the non-REM sleep where consciousness is increasingly lost. In contrast, the phase locking during REM sleep, when one dreams and retains consciousness, resembles that of the awake state.

Together, these data demonstrate that brain and heart dance together as if in a tango. As one tango dancer follows the other, the brain follows and aligns itself with the body's heartbeat. The converse is also true: the brain controls the heart through the autonomic nervous system and can speed up or slow down the heartbeat. Just as in a tango, the brain and the heart mutually align and synchronize with each other. This is what I call temporal (and ultimately temporo-spatial) alignment.

Stomach and Brain

How about the temporal alignment of the brain with organs other than the heart? In a recent study, Richter et al. (2017) investigated the relationship between the infraslow (around 0.05 Hz) rhythm generated by the stomach, as measured by a special device recording the stomach's movements, and the different frequencies in the brain's spontaneous activity, as measured by magnetoencephalography (MEG). They observed that the phase of the stomach's infraslow frequency (around 0.05 Hz) was coupled with the amplitude in the alpha range (10 to 11 Hz) of the brain's spontaneous activity. One can thus speak of cross-frequency coupling between body and brain, that is "gastro-cortical phase-amplitude coupling."

Neuronally, the gastro-cortical phase-amplitude coupling was associated with neural activity in two specific regions of the brain: the anterior insula and the occipital-parietal cortex. Richter et al. (2017) also measured the directionality of the coupling between the stomach and the heart. Using transfer entropy as measure, they show information transfer from the stomach to the brain and thus from the stomach's infraslow frequency phases to the brain's alpha amplitude in anterior insula and occipital cortex. In contrast, they did not observe reverse information transfer from the neural activity in the two brain regions to the stomach. This suggests that the brain aligns its alpha activity to the stomach's infraslow frequencies, rather than the reverse.

Taken together, these examples show that the brain aligns its own inner temporal structure to the temporal structure of the bodily organs' activity of heart and stomach. Such temporal alignment is possible through, for instance, actively shifting the phase onsets in the brain's neural activity – this is called "entrainment" (Lakatos et al. 2019). Temporal alignment of the brain to the body must thus be considered an active rather than passive process through which the brain's spontaneous activity can conform its own temporal structure to that of the body.

Time as the Common Currency of Body and Brain

How can two organs as different as the brain, on the one hand, and heart or stomach, on the other hand, communicate? For that, they must share a common currency. Imagine, for instance, that you have a friend from China who speaks Mandarin and not a word of French. Conversely, you speak French and not a word of Mandarin.

How can the two of you communicate? If you do not share a common language, you will remain unable to do so directly. You must share a common language, a third language beyond French and Mandarin. Though you are French and live on the other side of the channel, you learned English. At the same time, your Chinese friend also picked up plenty of English as it was taught to her at school. Accordingly, English is the shared language and thus the common currency that allows the two of you to communicate.

The situation between the two friends is analogous to the relationship between the brain and the stomach. The brain is featured by neuronal activity – the brain's language may thus be described as "neuronalese." The stomach, in contrast, is part of the gastro-intestinal tract characterized by various hormones that secret enzymes to digest food – one may want to speak of "gastrolese." For the brain's neuronalese, gastrolese is like Mandarin for a native French speaker. While for the stomach's gastrolese, neuronalese is like French for a native Mandarin speaker. No communication is possible as neither understands the other.

The situation becomes even more complicated when considering the heart. The heart may yet speak another language, "cardiacolese."

As all these languages differ from each other: brain, heart, and stomach can neither communicate with nor understand each other in the same way that you and your friend do not understand each other when speaking French and Mandarin. A common currency between brain, stomach, and heart is needed.

That is the moment where time comes in as it allows relating the different languages to each other. Based on their respective temporal features, the languages of heart, brain, and stomach are related to each other: the brain can sense the temporal features of the stomach's digestive processes – their changes over time – whereas it cannot understand gastrolese. The better the brain can sense the temporal features of the body's organs, the better it can align its own neuronal activity to them. Psychologically, this is manifest in more positive emotions, being more present within the own body, and, more generally, in better mental health.

From Brain-Body Time to the Contents of Consciousness

Consciousness: External Contents

Is the temporal alignment of the brain's temporal structure to the body's organs relevant for consciousness? So far, I have only demonstrated how both body and brain share a common currency through their temporal features. Does such body-brain coupling shape our consciousness? To address that question requires a turn to consciousness itself.

Consciousness can be characterized by different contents, including external contents from the environment and internal contents from the self. Both internal and external contents can be associated with consciousness. In both cases, the brain's temporal alignment with the body plays a central role, as the studies by the group around the French neuroscientist C. Tallon-Baudry demonstrate (see Tallon-Baudry et al. 2018 for an overview).

Let us start with external contents. In one study, Park et al. (2014) used MEG to investigate the impact of the heartbeat on conscious detection of visual stimuli. They investigated visual grating stimuli in a near-threshold way, that is, stimuli were presented at an intensity that was close to the individual subject's limit of conscious perception. While undergoing MEG and electrocardiogram recording, the subjects were exposed to these near-threshold visual stimuli and, for each, had to report whether or not they perceived and thus detected it.

The behavioural data showed a detection rate of 46 per cent, which indicates conscious perception of approximately half of the stimuli. Park et al. (2014) did not observe a direct relation of heartbeat and heartbeat variability with the subjects' detection rate. Hence, the heartbeat itself had no direct impact on conscious detection. That changed when Park et al. (2014) considered the neural correlates of how the heartbeats were processed in the brain, the heartbeat-evoked potential (HEP), as measured with MEG. The amplitude of the HEP predicted conscious detection: it was significantly higher for hits than misses. Accordingly, instead of the heartbeat itself, independent of its processing by the brain, the brain's processing of the heartbeat, i.e., the HEP, was central for associating consciousness to the visual stimuli.

The HEP, and its effects on conscious detection, was most predominantly located in anterior midline regions like the perigenual anterior cingulate cortex/ventromedial prefrontal cortex (PACC/

vmpfc). These regions are known to process interoceptive inputs from the body, including their integration with exteroceptive inputs from the environment. They also showed fluctuations in the spontaneous activity related to the HEP: the HEP differences between hits and misses during the detection task corresponded to HEP fluctuations as mediated by the spontaneous activity. This suggests a strongly temporal or better dynamic basis of heart-brain communication.

Together, these data suggest that the heartbeat affects and modulates the brain's spontaneous activity's spatiotemporal structure as measured by the HEP. That very same HEP-related modulation of the brain's spontaneous activity in turn shapes consciousness, that is, whether external contents become conscious. Accordingly, consciousness of external contents cannot be restricted to the brain alone. Nor can it be found in the body itself. Instead, the brain's temporal alignment with the body is central for associating consciousness with external contents.

Consciousness: Internal Contents

How about purely internal contents like the own self or autobiographical memories? A typical case where we withdraw from the external contents in our consciousness is when we no longer pay attention to the external environment. Instead, we follow our own internal thought contents. Where do these internal thought contents come from? This question is rather delicate. Neither the external world nor the body seem to play a role here. Instead, internal contents must be traced to the brain itself; they are the internal digestions of the brain independent of both the body and the world. The body may thus not be necessary for our consciousness of internal contents. That is not the case, though, as yet another study by the group around Tallon-Baudry shows.

Her group (Babo-Rebelo et al. 2016) tested whether consciousness of internal contents such as one's own self (in terms of "I" and "me") and its neural correlates in the brain's spontaneous activity (as measured with MEG) are coupled to the heartbeat. They again observed that spontaneous fluctuations of the HEP in PACC/VMPFC predicted the fluctuations in the consciousness of one's self in terms of either "I" (operationalized as "first-person perspective subject or agent of my own thoughts") or "me" (operationalized as "thinking about my own self").

Based on these data, Tallon-Baudry et al. (2018) postulate that the body is central for establishing the first-person perspective as hallmark feature of consciousness. By linking the internal and external contents' temporal processing in the brain to the dynamic of the body, the contents are connected with the person and their first-person perspective. That association, in turn, makes possible consciousness of both internal and external contents. Consciousness of both internal and external contents is thus ultimately based on the dynamic of brain-body coupling.

Consciousness: Kant's Errors

The question for the neural basis of internal and external contents of consciousness is not a trivial one. Philosophers were split about this. The empiricist philosopher David Hume rejects any internal contents in consciousness; consciousness, according to him, is only determined by external contents from the environment. Against that, the idealist philosopher Immanuel Kant claims that our consciousness does indeed possess internal contents (Northoff 2012a; Northoff 2012b).

That split between internal and external contents appears within current-day neuroscience. External contents are related to the sensory regions of the brain while internal contents are associated with higher-order regions like the prefrontal cortex and the default-mode network (DMN). This is well-reflected in the currently predominate dual model of cognition (DMC) (Northoff et al. 2022): the DMC associates lower-order regions featured by task-related activity with external content, while the DMN in the resting state is related to internal content.

However, despite their difference in origin, both internal and external contents share some commonality: both can be associated with consciousness. On the neuronal side, the spontaneous activity featured by its dynamic provides the baseline, that is reference or standard, for both internal and external content. This amounts to a baseline model of cognition (Northoff et al. 2022).

To deny that our consciousness shows internal contents, as Hume did, is to neglect one core feature of consciousness, namely our internal perceptual and thought contents. On the other hand, their clear separation, as Kant assumed, is also not feasible either, as the contents of our consciousness are all based, at least in part, on the brain's spontaneous activity and its dynamic. Neither of these solutions holds when looking at the data.

The data agree with Kant that there are both internal and external contents and they are processed by different regions in the brain. Moreover, Kant argued, correctly, that our consciousness shows both internal and external contents. Importantly though, the data go beyond Kant. Kant assumed two different forms of consciousness: internal and external. Yet this is not empirically plausible, as there are internal and external contents but not two forms of consciousness. All forms of consciousness are based upon the brain's spontaneous activity and its dynamic: the data show that the brain-heart relation mechanism underlies the association that both internal and external contents have with consciousness. That extends beyond Kant, who limited especially internal consciousness to the boundaries or confines of the mind without considering the body let alone the world.

In a nutshell, Kant was ingenious enough to see that our consciousness shows both internal and external contents. However, he remained unable to consider that, despite their differences, their association with consciousness can be traced to one and the same underlying mechanism. He was unable to take into view that since, as I assume, he limited his focus to the mind or, as we would say nowadays, to the brain. That prevented him from taking into view what lies beyond the brain, that is, the brain's tango with both body and world.

Is Consciousness Special or Non-special?

Consciousness: Dualism of Mind and Body

Consciousness is the core of our mental life. We experience the book in front us. It elicits a certain quality in our subjective experience, the redness of the colour red – qualia as philosophers like Thomas Nagel say. He wanted to know how the bat experiences the world. He concluded that we cannot know what it is like for the bat to experience the world. We, as humans, remain unable to take on the perspective of the bat or the first-bat perspective, as analogous to our first-person perspective. Some philosophers like Colin McGinn

(McGinn 1991) therefore conclude that we, in principle, remain unable to explain consciousness which thus must remain mysterious to us. Does this mean that any scientific investigation of consciousness is futile?

Philosophers like René Descartes in the sixteenth century considered consciousness as special. At that time, science took off, especially physics. Time was characterized by distinct points as those can be observed in an objective way from the outside, i.e., in third-person perspective. The body could be investigated in this way. In contrast, the mind and especially consciousness defied such view. It can neither be observed from the outside in an objective way nor be captured in third-person perspective. Instead, consciousness is an experience inside of ourselves that is essentially subjective and can only be accounted for in first-person perspective. Consciousness must thus be special when compared to the body. Descartes concludes that there must be two different properties or substances in the world, a physical one and a mental one, in what is known as mind-body dualism.

We nowadays reject the mind-body dualism of Descartes. Due to increasing insight into the brain, we now know that consciousness is related to the brain. Philosophers and neuroscientists alike claim that there is no such special substance like a mind. Instead, the mind is the brain, or, by extension, is the interaction of brain and body as assumed in the embodied (and extended) approaches to the mind. The mind-body dualism of Descartes and others is replaced by the question for the neural correlates of consciousness (NCC). The NCC refer to those neuronal features in the brain that are sufficient to yield consciousness. However, amidst all the enthusiasm for the NCC, I claim that our current search for it is missing something fundamental.

Consciousness Is Special: "Old Wine in New Bottles"

Consciousness is special even in our times. Unlike in the times of Descartes, the specialness of consciousness is no longer located outside of the brain in a mind or mental properties. Instead, the specialness of consciousness is now inside the brain itself – consciousness is supposed to be mediated by a special region or mechanism within the brain as distinguished from the other non-special regions and mechanisms. In short, neuronal specialness replaces mental specialness.

The neuronal specialness of consciousness provides the background assumption for most of the major neuroscientific theories of consciousness. Special neuronal mechanisms are supposed to consist in integration of information (integrated information theory [IIT]) (Tononi et al. 2016), access to a global neuronal workspace by the brain (global neuronal workspace theory [GNWT]) (Dehaene et al. 2014; Dehaene et al. 2017; Dehaene and Changeux 2011, recurrent processing through feedback loops (Lamme 2018), higher-order processing (Lau and Rosemnthal 2011), or predictive coding (Friston 2010) (see also Mashour et al. 2020; Northoff and Lamme 2020; Seth and Bayne 2022 for overviews).

Despite their differences, these theories share the mostly implicit assumption of a special neuronal mechanism for consciousness as distinct from all other neuronal mechanisms in the brain. Given such neuronal specialness, the current neuroscientific theories of consciousness stand in a dualistic tradition that can be traced to Descartes. They are nothing but "old wine in new bottles": replacing the mind, the brain provides a new bottle for consciousness within which it is still conceived as special and thus as old wine.

Consciousness Is Non-special: "New Wine in New Bottles"

How can we overcome the neuronal specialness of consciousness? Rather than supposing special features, we would need to assume the most basic features of the brain, the body, and even the world to mediate consciousness. This leads us back to time. Time is considered by many physicists and philosophers to be the most fundamental and basic constituents of the world. In that case, time should also shape the brain including its relation to the world, the world-brain relation. As discussed in the previous chapter, the dynamic of the world-brain relationship is key for consciousness as it situates and embeds us within the world. Consciousness may thus be based on time (and space) in the brain, the body, and the world – this is the key claim of the temporo-spatial theory of consciousness (TTC) (Northoff and Huang 2017; Northoff 2013; Northoff 2014b; Northoff 2016; Northoff 2018; Northoff and Zilio 2022a; Northoff and Zilio 2022b).

The TTC is an integrated philosophical and neuroscientific theory of consciousness. One key claim is that the construction of time (and space) is the common currency of the world, the brain, and the consciousness (Northoff et al. 2020a; Northoff et al. 2020b). The world constructs its inner time which, in part, is shared by the brain's construction of its own inner time which, in turn, shares it with consciousness. Consciousness must be conceived as essentially temporal and thus dynamic. Since the world features the same dynamic as both the brain and the consciousness, the latter is not special at all. Such non-specialness is the core assumption of the TTC, which distinguishes it from all other consciousness theories (Northoff and Lamme 2020).

The TTC assumes that the continuous presence of consciousness is based and dependent on the continuous change of the brain's neuronal activity including its relationship to the body and the world. Continuous change on the neuronal level of the brain makes possible the continuous presence of consciousness. Change and presence are no longer mutually exclusive, as assumed in both past philosophy and present-day neuroscience. Instead, the continuous presence of consciousness is dependent upon the continuous change of the brain's neuronal activity. Both neural and mental activity are thus intimately connected through their dynamic, that is, the continuity over time.

In sum, rather than being special within the brain and the world, consciousness is not special at all. Instead, it is the non-specialness of consciousness, its reliance on the most basic feature of the world –time – that even makes consciousness possible. This is what the empirical data tell us. We thus do not only provide "new bottles," i.e., empirical and neuronal framing of consciousness, but also pour "new wine," i.e., the temporal or dynamic basis of consciousness across the distinction of brain, body, and world, into them.

Conclusion

Are the body and its time relevant for consciousness? Yes. Our brain's time needs to align itself to the world time through our body's time for consciousness to be possible. This is manifest in our consciousness, which enables us to experience part of the wider world rather than being restricted to either the brain or the body. In the words of the famous neuroscientist Gerald Edelman (1929–2014), a "behavioral trinity of brain, body, and world" is central for consciousness (Edelman et al. 2011, 4). Consciousness is embodied, embedded, and enacted; it extends beyond the boundaries of the brain to the body and the world.

One can characterize the relation of the body and the brain by double directionality. There is directionality from the body to the brain, i.e., body-brain relation – the body impacts the brain's interoceptive functions with the latter actively aligning themselves to the former in a dynamic way in what is known as temporal alignment. The bodywaves shape the brainwaves – this is key for consciousness. At the same time, there is directionality from the brain to the body, i.e., brain-body relation – the brain impacts the body and can regulate it according to its own temporal structure. To return once more to the tango: the brain's inner time dances tango with the body's inner time. Though sometimes they are in sync while at other moments they are not, their tango is key for shaping the contents, internal and external, of consciousness.

4
Self Time and Its Duration

Introduction

Self: Change and Persistence

Who is the surfer on the surfboard gliding on the ocean's waves? The surfer is the person whose self stands on the surfboard and directs it through the waves. That self is the topic in the present chapter.

The self is always there. Except for certain extreme states that may require medical intervention (see chapter 5), we never lose our sense of self. We develop our sense of self from an early age onwards, and it lasts basically until we die. Most remarkably, our sense of self persists throughout the bodily and environmental changes we experience in life. Our body grows from a child into an adult and continues to age until we become old, fragile, and wrinkled. The environment changes with major social, cultural, and political fluctuations throughout our human lifespan. Our thoughts and emotions shift over time.

Despite these physical and mental changes, we experience or sense ourselves as one and the same self over time. I have always sensed myself as the self of Georg Northoff even though my body has aged, I now live in Canada rather than my original home country of Germany, and my emotions and cognitions have shifted over time. Others remain the same even if they switch from being a left-wing communist in their youth to a right-wing fascist in adulthood. Despite all the changes, we nevertheless experience ourselves as one and the same self. This amounts to what I call the "time paradox of self."

The Time Paradox of Self: Co-occurrence of Sameness and Difference

The time paradox of self indicates that the self combines both sameness and difference. Our self is different all the time, as reflected in its continuous bodily and mental changes. Yet our self and how we experience it – our sense of self – remains the same throughout all those changes. Hence, our self seems to combine sameness and difference at once.

The philosopher may want to jump in and say that sameness and difference are not compatible on purely logical grounds: sameness excludes difference just as difference excludes sameness. To combine both sameness and difference is thus impossible. However, such a combination seems to be exactly the case when it comes to the self. Even if logically impossible, there is no doubt that the self exists and is real, at least in our experience.

How can the self be based on two mutually exclusive and incompatible features like sameness and difference? The answer to this question leads us again back to time. Difference presupposes the flow of time: continuous change with flow from the past over the present to the future. Sameness, in contrast, presupposes the stillstand of time: no change, no flow. To account for the self, we thus need to combine flow and stillstand. That leads us to the notion of duration. Duration refers to the time of our inner subjective experience, the lived time. It includes the experience of both sameness and difference and thereby of persistence and change. How is this duration constituted? This leads us to the inner time of our brain and how it mediates our sense of self.

The Brain's Inner Time and the Self

Spatial Structure of the Self in the Brain: Cortical Midline Structures and the Mental Self

Before getting into the temporal features of the brain and the self, we need to understand how the self is "located" in the brain. Cortical midline structures (CMS) like the perigenual anterior cingulate cortex (PACC) and posterior cingulate cortex (PCC), as well as other regions inside and outside the cortical midline structures, are most consistently activated during self-related processing. Though the PACC and PCC (and other midline regions like the dorsomedial prefrontal cortex, supragenual anterior cingulate cortex, and medial parietal cortex) are related to differential aspects of self-related processing, they are most often conjointly recruited and activated (to different degrees) during different degrees and aspects of self-related processing.

Interestingly, the data show significant neural overlap between the high resting state and self-related activity levels in the PACC and PCC. Several studies observed that self-specific stimuli did not induce activity change in the PACC and PCC during task-evoked activity when compared to their resting state activity levels; such "rest-self overlap" (Bai et al. 2015) was further confirmed by a meta-analysis showing the PACC and PCC as overlapping regions during both resting state and self-related processing (Qin and Northoff 2011).

One may even go one step further and show that resting state activity and pre-stimulus activity levels predict the degree of self-consciousness. This is indeed true as the temporal features of both resting state activity (Wolff et al. 2019; Huang et al. 2016; Kolvoort et al. 2020; Smith et al. 2022) and prestimulus activity (Bai et al. 2015) predict the degree to which one is aware of being a self with certain psychological features or self-specificity assigned to subsequent stimuli (see below). These findings of rest-self prediction imply that the resting state itself encodes or contains some information about our self in yet unclear ways. The assumptions of "rest-self overlap" may thus be accompanied by the one of "rest-self containment" (Northoff 2016) or "self-representation" (Sui and Humphreys 2016, 4).

Finally, a recent meta-analysis by Qin et al. (2020) shows that the PACC and PCC are mainly related to the mental self, or the awareness and experience of one's self as self. That is distinguished from the exteroceptive-proprioceptive self that refers to the own external bodily boundaries; this involves other regions like the temporo-parietal junction and premotor cortex. Finally, the lowest layer of self consists in the interoceptive self, the experience of one's inner body; that recruits insula, dorsal anterior cingulate cortex, and subcortical regions like the thalamus.

How are the three layers of self – mental, interoceptive, and exteroceptive – related to each other? Qin et al. (2020) show that the lower regions of the interoceptive self are also recruited by both the exteroceptive and mental self, while the exteroceptive self's regions are co-activated with the PACC and PCC by the mental self. This amounts to a three-layer organization and ultimately topography of the self, with the three layers nested within each other like Russian dolls. The biggest Russian doll usually dominates our view. Analogously, the self with the most recruited regions – the mental self – dominates the experience of our self. Therefore, my focus below is on the mental self with the PACC and PCC as key regions while, for the sake of simplicity and clarity, I more or less neglect intero- and exteroceptive self.

The Inner Time of Our Brain: Big and Powerful Waves in Cortical Midline Structures

I have so far focused only on the spatial side of the self. How about the temporal side? This points to the spontaneous activity's inner time. I pointed out that the spontaneous activity exhibits and constructs a complex temporal structure with temporal duration (see chapters 1 and 3). Such temporal duration is manifest in features like autocorrelation, cross-frequency coupling, and scale-free activity (see chapters 1 and 3). I now investigate whether these features are related to our sense of self.

Zirui Huang, a Chinese postdoctoral student in my group at the time, investigated how the spontaneous activity as measured in fMRI is related to our sense of self (Huang et al. 2016). fMRI can measure the infraslow frequency range between 0.01 and 0.10 Hz which shows much power and follows scale-free distribution (with slower frequencies exhibiting more power than faster ones). Specifically, he measured scale-free activity using the power law exponent (PLE) in the spontaneous activity of the two key regions of the CMS: the PACC and PCC.

He observed that the PACC and PCC exhibit the highest PLE values in their spontaneous activity when compared to all other regions in the brain. Unlike all other regions, these two show the strongest power in the slower frequencies and relatively weaker power in the faster ones. One can compare that to the occurrence of waves in the ocean. When you sit in front of the ocean, you can observe waves of different speed and power. Faster waves are more frequent, usually small, and less powerful. Slower waves come less often, are usually big, and can be extremely powerful as they may swipe away all your belongings at the beach.

We face an analogous scenario in the brain. The cortical midline structures like the PACC and PCC exhibit the strongest waves, slow but powerful and big in amplitude – they exert the real force behind the many smaller ones. In contrast, regions outside the PACC and PCC provide smaller waves that are faster, lower in amplitude, and less powerful. Due to their slow but powerful waves, the inner time of the PACC and PCC is different than the inner time of the other structures: the former show longer duration with more elapsed time than the latter.

The Inner Time of Our Self: Big and Powerful Waves

How can we measure the self? One way is to assess a subject's sense of self by employing a psychological scale like the self-consciousness scale (SCS). The SCS is a questionnaire where subjects are asked about their self across private (like "I am often in my own inner thoughts"), public ("I am an outgoing person"), and social ("I like to connect with other people") dimensions. One can then relate the individual subjects' scores of the SCS to their spontaneous activity as measured in fMRI. This is exactly what Huang did.

Huang observed a direct relationship of the psychological scores of the self-consciousness scale with the waves of the brain's spontaneous activity in the PACC and PCC. Specifically, the degree of scale-free activity, as measured by the PLE, directly correlated with the degree of private self-consciousness: the higher the PLE in the spontaneous activity's PACC and PCC, the higher the degree of the respective subject's private self-consciousness.

fMRI covers a frequency range of 0.01 to 0.10 Hz, which covers durations of one hundred seconds to ten seconds. However, there are also faster frequencies with shorter durations which can be measured by EEG. Wolff et al. (2019) applied the same self-consciousness scale to subjects in EEG and observed the same relationship as in fMRI: more powerful slower frequencies (relative to faster ones) relate to higher sense of self (e.g., private self-consciousness). Together, these results suggest that the self operates across all frequencies, including both slower and faster ones, in the same way. In other words, the more powerful slower frequencies strongly shape our sense of self.

From Inner Time to Duration

Duration of the Brain: Extension of Inner Time through the Slower Frequencies' Cycle Durations

What do these results mean? They mean that stronger power in the brain's slower waves relative to its faster ones yields a stronger sense of self. Accordingly, our sense of self is strongly shaped by the energy or power of the brain's slower and more powerful frequencies, including the degree to which they structure its faster and less powerful frequencies. How do the strong powerful waves of the brain constitute the inner time of the self, that is, its duration?

Waves are about temporal duration. Stronger power in the slower frequencies means that their waves are not only more powerful but also last much longer, with longer cycle durations. For instance, the cycle duration of the frequency 0.01 Hz is one hundred seconds while the one of 0.10 Hz lasts only ten seconds. The self is related to the power of the slower frequencies, which have longer cycle durations that entail longer temporal duration. Therefore we can say that our self is about temporal duration: the stronger the power of the slower frequencies, the longer and more extended their temporal duration, and the stronger our sense of self.

Duration on the level of neuronal activity translates into a stronger sense of self – the latter thereby shows by itself an inner duration. How can we test and investigate that? If the self shows longer duration, we would expect it to integrate temporally distinct stimuli over longer periods of time. This was tested by another student of mine, Ivar Kolvoort from the Netherlands.

Kolvoort investigated whether we can associate self-specificity to external stimuli over longer periods of time (delays) than non-self-specificity. He showed that we can assign self-specificity to external stimuli over longer delay periods than non-self-specific stimuli. Even more important, this was directly related to the degree of the spontaneous activity's PLE in PACC and PCC: the more powerful the slower frequencies with their longer cycle durations, the longer the temporal delays over which self-specificity could be maintained (Kolvoort et al. 2020). This again suggests that the extension of the brain's inner time through the slower frequencies' phase cycle durations is key for our sense of self.

Duration of the Self: "Common Currency of Self and Brain"

These data strongly support the assumption that our self relates to time, and specifically to duration. Longer duration is related to a more extended inner time, and the greater that extension, the stronger our sense of self. Duration as the extension of inner time is mediated by our brain's inner time: neural activity can extend its inner time by its own inner timescales like the longer phase cycle durations of its slower frequencies. The brain thus constructs its own inner time extensions on the neuronal level, and these are manifest on the psychological level of self, that is, in our experience of a temporally more continuous and thereby stronger sense of self.

How can we further support this assumption? Duration as the extension of inner time entails temporal continuity. Temporal continuity can be measured by the degree to which the activity at one point in time relates to the degrees of activity at subsequent time points; this is formalized as the autocorrelation window (ACW). Applying EEG, David Smith from my group (Smith et al. 2022) showed that the brain's degree of temporal continuity, as measured by the ACW, is significantly longer during tasks related to the self than those not related to the self (e.g., listening to one's own and others' stories over eight minutes), as well as longer than in the resting state.

Together, these data show that the brain extends its timescales when the self is involved. Our self thus likes slower but more powerful frequencies with their longer phase cycles and timescales – our self's inner time is based on the degree to which the brain can extend its own inner time. Accordingly, duration as the extension of inner time is manifest on both levels, neural and psychological – the brain and the self share it as their common currency (Northoff et al. 2020a; Northoff et al. 2020b).

Duration of Personal Identity: Scale-Freeness across Shorter and Longer Timescales

Duration extends beyond the timescales of fMRI and EEG. Philosophers may for instance want to argue that the self entails a strong diachronic component, namely personal identity, which holds over a whole lifetime. The timescale of personal identity implies a span of the potential decades within a person's lifetime, whereas the timescale of measuring with PLE in fMRI in a frequency range of about 0.01 Hz implies a span of about one hundred seconds. Measuring fMRI continuously throughout a whole lifetime remains impossible, obviously.

However, we can assume that shorter and longer timescales are intimately related to each other: following their scale-free character, each shorter and less powerful timescale nests within the next longer and more powerful one. In other terms, the timescales of fMRI may be more or less self-similar to the much longer ones of a whole lifetime. Given such scale-freeness, we may assume that the PLE measured in fMRI including its relation to the self may reflect upon, to certain degree, the much longer timescales of a lifetime, namely those of personal identity. We may consequently assume that subjects who show stronger PLE in their brain's spontaneous activity exhibit not only higher private self-consciousness but also stronger personal identity.

The Self: Atemporal or Temporal?

Logical World of Philosophy: The Self Is Atemporal

There is an extensive discussion about the nature of self in philosophy. Philosophers especially point out the sameness and persistence of the self, for which reason the self has often been viewed as a specific property or substance. For instance, René Descartes (1595–1650) views a substance or a property as enduring and therefore atemporal. He (and others) conceived this atemporal nature as necessary for something like the self to remain the same and persistent over time. One may distinguish physical and mental substances or properties. For instance, our body remains the same and is persistent and therefore can be considered a physical substance or property. On the mental level, the self is persistent and may therefore be conceived as mental substance or property. Despite their differences, both physical and mental properties share an atemporal nature, that is, the enduring character without any change.

Is there any evidence for a mental substance or property as basis of the self? Our data show that the self is related to the brain and its physical features, that is, its temporo-spatial dynamic. Hence, the self may not be mental but physical – it is certainly not a mental substance or property as distinguished from the body as physical substance or property (see also Churchland 2002; Metzinger 2003). That does not exclude the possibility, though, that the self is a physical substance or property. In that case, one would assume that the persistence and sameness of the self is based on the atemporal nature of such physical substance or property. Accordingly, the self would be atemporal in the purely logical world of philosophy.

Biological or Natural World of Neuroscience: The Self Is Temporal

Is the self indeed atemporal? Empirical evidence suggests otherwise. The data clearly show that our sense of self is intimately connected with the brain's inner time, that is, its degree of extension through the slow frequencies' phase cycles and timescales. The duration of the brain's inner time, in turn, translates into the duration of the self's inner time. The self is thus intrinsically temporal rather than atemporal. This speaks against the assumption of any kind of property, whether physical or mental, underlying the self.

How, then, can the inner time and its duration account for the persistence and sameness of self over time? Philosophers mistakenly assume that an atemporal property or substance accounts for the persistence and sameness of self, and deem any kind of temporality with change and difference to stand opposite to persistence and sameness. While that may be true from a purely logical point of view, it does not hold empirically. We can see that the brain's inner time is characterized by both persistence/sameness and change/difference: the long cycle duration of the slower frequencies with their long timescales relate to sameness and persistence, while the faster frequencies with their shorter timescales relate to change and difference. Even stronger, the data suggest intimate relationship with interdependence of slower and faster frequencies such that the former structure and organize the latter.

Together, these empirical observations strongly suggest co-occurrence and interdependence of change/difference and persistence/sameness within the duration of the brain's inner time. Importantly, such duration of inner time, i.e., its extension, is manifest in both the brain's neural activity and our sense of self as their common currency. We experience our self as both changing and persistent and therefore as different and the same simultaneously. The empirical data strongly suggest that this can be traced to the degree to which the brain's inner time extends through its phase cycles and timescales. Hence, my answer to the early philosopher is as follows: though persistence/sameness and change/difference are incompatible in the context of a purely logical world, they are well compatible in the context of a natural world such as that of the biological world of brain and self.

What Is Duration? Henri Bergson and the Notion of Elapsed Time

The self is duration. But to what does "duration" refer? Duration describes the amount of elapsed time between two events. In the context of music, duration describes the amount of time between the beginning and ending of a note, a phrase or melody, a piece, etc. It can be specified by beat and meter – the rhythm is an essential ingredient of duration. This implies two perspectives on duration: outer and inner.

From an outer, or third-person, perspective, we perceive the beginning and end of a phrase or melody. In contrast, we do not perceive and observe the inner duration of the event itself, that is, how long it stretches and elapses over time – that we can only measure in an indirect way, namely by a meter of time like the clock. Accordingly, measurement of an event in outer time only accounts for change and thus difference – the beginning of a tone or phrase is contrasted with its end.

In contrast, such objective measurement from the outer perspective does not account for the sameness of the event over time, that is, its duration. The French philosopher Henri Bergson (1859–1941) at the beginning of the twentieth century points out the key importance of duration (Bergson 1946). Duration is the inner time of the event itself independent of our outer observation and measurement of the time of that very same event by an external clock. How can we consider the duration of an event?

Bergson argues that we need to take an inner (rather than outer) view or perspective from within the event itself – this allows us to apprehend the elapsed time itself which is what he describes as duration:

Let us imagine an infinitely small piece of elastic, contracted, if that were possible, to a mathematical point. Let us draw it out gradually in such a way as to bring out of the point a line which will grow progressively longer. Let us fix our attention not on the line as line, but on the action which traces it. Let us consider that this action, in spite of its duration, is indivisible if one supposes that it goes on without stopping; that, if we intercalate a stop in it, we make two actions of it instead of one and that each of these actions will then be the indivisible of which we speak; that it is not the moving act itself which is never indivisible, but the motionless line it lays down beneath it like a track in space. Let us take our mind off the space subtending the movement and concentrate solely on the movement itself, on the act of tension or extension, in short, on pure mobility. This time we shall have a more exact image of our development in duration. (Bergson 1946, 164–5)

Duration Constitutes Our Self

Following Bergson, duration combines sameness and difference. One event remains the same through the time it occurs, that is, from the beginning to the end – duration entails sameness, which we perceive as persistence. At the same time, duration describes the elapsed time from the beginning to the end of the event – this entails difference, namely the difference between the presence and absence of the event yielding what we describe as change. Duration as the amount of elapsed time thus does not only combine sameness and difference over time, but also makes possible the co-occurrence of change and persistence in time.

All that may sound theoretical and abstract, but it is exactly the way your own self can be characterized. The self is duration, it is elapsed time, which can be conceived as extension of our inner time. It combines both sameness and difference simultaneously: our self remains the same and persists throughout all its bodily, environmental, and mental differences yielded by the physical and mental changes it undergoes over our lifetime. All that is possible by extending our own inner time through the brain's extension of its inner time by the slower frequencies' phase cycle durations and longer timescales.

Conclusion

What is the self? The self features a time paradox: it combines both change/difference and persistence/sameness, which are logically exclusive and thus incompatible with each other. However, the experience of self tells a different story. We experience continuous change and difference of our self, which nevertheless remains persistent and thus the same over time. Philosophers assumed that an enduring atemporal substance or property underlies the sameness/persistence of self, and thus conceive of the self as atemporal. That, however, leaves open the change/difference of self.

The empirical data strongly support the co-occurrence of both change/difference and persistence/sameness within the brain's inner time. The brain's inner time can take on different degrees of extension through the durations of especially its slow frequencies' phase cycle and longer timescales. These degrees of extension of the brain's inner time, in turn, are manifest on the psychological level of self. The self is therefore intrinsically temporal rather than atemporal. This allows it to combine the seemingly paradoxical features of change/difference and persistence/sameness within the extension of its inner time, the elapsed time as Bergson notes. Accordingly, duration featured by extension of inner time resolves the time paradox of the self.

Being featured by duration as the extension of inner time, the self can be compared to the surfer on the surfboard. The surfer remains the same but, with every wave, takes on a different position on their surfboard to stay afloat. The surfer thus connects sameness and difference. Moreover, the surfer is part of the wave itself and thus has an inner perspective on the dynamics of the wave itself from their inside or inner view of the wave. At the same time, the surfer stands above the wave and can thus take on an outer perspective. Finally, the surfer combines change, in the form of the waves, and persistence, by remaining on the surfboard despite the change.

Accordingly, in the same way we characterize surfing by duration as elapsed time, we can describe our self as a mental surfer who uses consciousness as surfboard to ride and surf on the waves of the world. Surfing is about duration, that is, the elapsed time between the beginning and end of one wave. The same applies to our self who is continuously surfing on the various waves of the world. I thus characterize the self by the degree of extension of its inner time, that is, duration as the elapsed time between different waves in the world as mediated by our brain's waves.

5
Time Speed in Brain and Mind

Introduction

Speed of Time – Speed of the Brain

I have covered various features of time including waves, synchronization, change, and duration. What about speed of time? The speed of an object is defined by the magnitude of its velocity that describes the change in position or the covered distance. For instance, in horse racing, the horse that changes its position faster than the other horses will win the race. We can measure speed by dividing the distance raced by the horse divided by the duration of the time interval – the winning horse thus covers the most distance within the shortest interval of elapsed time, the duration (see chapter 4).

Speed is usually considered an objective quantity. We can measure the speed of horses, cars, runners, and so. One would usually not associate speed with the brain. The brain seems to possess no speed at all. It does not change position and exhibits no velocity – speed, as calculated by distance travelled divided by duration, should thus amount to zero. To assume otherwise would be counterintuitive if not absurd.

However, the opposite is true. The brain's spontaneous activity constructs its own speed as the change in position or covered distance within an elapsed time interval. The brain's spontaneous activity exhibits different frequencies: slow ones that are powerful and faster ones that are less powerful. The slow frequencies exhibit long cycle durations, while the faster ones exhibit short cycle durations. The long cycle durations are more sluggish to change their position and therefore require more time in which to do so or to travel a certain distance compared to the shorter cycle durations. That makes it clear why, as in their names, slow and fast frequencies can be characterized by different speeds.

From Horse Racing to the Brain's Different Frequencies

Given that the brain exhibits different frequencies on a range from very slow to very fast, the brain cannot be characterized by one speed but by several co-occurring speeds. Compare that to horse racing. There is not only one horse on the course but several horses who race against each other. There are very slow horses who lag behind, medium slow or fast horses in the middle of the pack, and some very fast horses that are at the top and set the pace for the rest of the field.

Something is wrong in this comparison, though. The slow frequencies lag behind the fast ones in their speed, yes. However, the slow frequencies are much more powerful than the fast ones. Applied to horse racing, this means that the slow horses may lag behind at the beginning of the race given that their velocity is lower than that of the fast horses. However, given the powerful nature of the slow frequencies, the horses with strong slow frequencies may ultimately overtake those horses exhibiting more fast but less powerful frequencies.

Accordingly, at the beginning of the race, the horses with more faster frequencies will lead. However, over time with increasing distance, the horse with stronger slow frequencies will overtake the others as the slower frequencies are ultimately more powerful than the faster ones. Hence, if the race is short, the horse with stronger faster frequencies might win. If the race is long, the horse with stronger slower frequencies will ultimately pass the others and take on the lead until the finish line. Accordingly, it is the balance of slow-fast frequencies of the horse itself and the external demands, the length of the course, that make the horse a winner or loser.

The same occurs in the brain. During short moments or durations of time, faster frequencies may take on the lead as their cycle durations are suited for short distances. However, during longer moments or durations of time, slower frequencies take over with their long cycle durations, allowing changes in position over longer distances. The co-occurrence of slow and fast frequencies in the brain's neuronal activity can thus be compared to a horse race: the brain's different frequencies compete like different horses in a race.

From Neuronal Speed to Mental Speed

Why is the speed of the brain relevant for mental features? I have demonstrated that the brain's construction of the brain's inner time is central for constituting mental features like consciousness and sense of self. The same holds in the case of speed of time. The brain's speed of time in its neuronal activity translates into the speed of time in our consciousness – neuronal speed is mental speed. The brain's speed translates into behaviour.

Different individual subjects exhibit different speeds. People who are slower in their external reactions usually have stronger internal thoughts. A typical example are philosophers who, before saying anything and acting towards the external world, must first think and dwell in their own inner thoughts. Conversely, soccer players must be fast with acting and reacting quickly to the ball without much thinking as otherwise they would miss the ball. What is good for the philosopher, the slow speed, is bad for the soccer player. Conversely, what is good for the soccer player, the fast speed, is disastrous for the philosopher.

Investigations in the brain show that its sensory regions, where external stimuli are processed, are much faster in their neural activity with more power in the faster frequencies compared to the slower ones. In contrast, if we dwell more in our own inner thoughts and let our mind wander, our midline regions as part of the default-mode network (DMN) become more active – these regions show much more power in the slower frequencies when compared relatively to the power in the faster ones. Put together, the brain's external orientation towards the outer world is mediated by faster frequencies while its inward orientation towards the own inner thoughts is related to slower frequencies. Briefly, external is fast, internal is slow. This can be taken to extremes in mood disorders like depression and mania.

Too Slow or Too Fast: Depression and Mania

Mania and Depression: Inner and Outer Time Speed

One central element of consciousness is our experience or perception of time, as reflected in what William James (1890) describes as "stream of consciousness." In addition to projecting ourselves into both future and past, perception of the speed of time is central in our stream of consciousness. Even if an event unfolds in an objectively slow speed, we may nevertheless perceive it as fast if we are fascinated by or drawn into the event – afterwards, we say "time flies," meaning that our subjective time speed perception was faster (with a perceived shorter duration of the event) than the objective physical time duration of the event itself. The opposite is the case if we are bored; in that case, time does not move at all in our subjective perception so that we will perceive the duration of the event much slower and longer than it actually was.

Time speed perception is highly atypical in depression and mania, albeit in opposite ways. Psychiatrists like Thomas Fuchs (2013) distinguish between inner and outer time speed perception. Inner time speed perception describes how we perceive the speed of our self as slow or fast, while outer time speed perception describes how we perceive the speed of events or objects in the outer world. In subjects without depression or mania, both inner and outer time speed perception are more or less balanced and thus synchronized.

In contrast, inner and outer time speed perception are desynchronized in depression and mania. Patients with depression show an atypically slow inner time speed perception – "nothing changes for them," which results in the perception of a "standstill of time." They also show an atypically fast outer time speed perception where they experience events and objects in the outer world as too fast and rapidly changing, even if they are objectively slow, and as a result feel easily stressed when confronted with external demands. The opposite pattern can be observed in patients with mania, who perceive their inner time and self as extremely fast and the events and objects in the outer world as too slow. As in depression, there is disharmony between the inner and outer time perception, though in the opposite direction.

How to Measure the Time Speed of the Brain: Neuronal Variability

Does time speed on the neuronal level of the brain correspond to the mental speed? Studies in healthy subjects show the involvement of regions in the somatomotor network (SMN), including the supplementary motor area (SMA); premotor cortex; medial and superior frontal gyrus; inferior parietal cortex; posterior insula; and subcortical regions like thalamus, pallidum, and putamen during tasks that require inner time speed perception. Since these regions are apparently involved in the perception of inner time speed, they have been described as a "neural timing circuit." In contrast, primary sensory cortices like the visual cortex or visual network (VN) may mediate outer time speed perception, as they provide the first encounter of the external stimuli with the subject's brain.

The next step is to determine a neuronal measure of time speed perception. Time speed perception is about change: if there is little change, time speed is perceived as slow (as in depression), while if there is a lot of change, time speed is perceived as fast (as in mania). Therefore, we may want to look for a measure of change on the neuronal level to account for time speed on the perceptual level. Such neuronal measure of change consists in neuronal variability, which measures the degree of change or variance in the amplitude of the signal from time point to time point as operationalized by the standard deviation (SD).

This leads to specific hypotheses. If there is a high degree of SD in SMN indexing much change in the amplitude of the spontaneous activity, then there is a high degree of neuronal speed which should translate into a subjectively perceived fast inner time speed. If, in contrast, there is not much neuronal change in SMN, SD is low, leading to the perception of slow inner time speed. The same should hold analogously for SD in VN mediating outer time speed perception.

Neuronal Variability in Depression and Mania

The results my colleagues and I discovered in depression and mania confirm these hypotheses (see Northoff et al. 2018). As expected, depressed bipolar patients showed decreased SD in SMN as accounting for their perception of atypically slow inner time speed. At the same time, their SD in VN was atypically high, in accordance with their perception of an atypical fast outer time speed in the external world. The perceptual disbalance between inner and outer time speed in depressed patients may thus be related to a neuronal disbalance between SD in SMN and SD in VN.

As on the perceptual level, manic patients showed an opposite pattern on the neuronal level. They exhibited an abnormally high SD in SMN mirroring their perception of an abnormally fast inner time speed. While their SD in VN was rather low, accounting for their perception of an abnormally slow external time speed (see Northoff et al. 2018). Accordingly, like depressed patients, manic patients show a disbalance on both levels neuronal and perceptual, albeit in an opposite way.

We also correlated the psychopathological symptoms with SD in SMN, VN, and the SMN-VN SD balance. This yielded a significant correlation of psychopathological symptoms, with SMN-VN SD balance but not with the SD in the single networks themselves. That strongly suggests that it is the balance between the two networks' SD, rather than the single networks' SD themselves, that is central here. The desynchronization between inner and outer time speed perception can thus be traced to a corresponding disbalance in SD between SMN and VN. We therefore assume the SMN-VN network's disbalance to be the neuronal mechanism that underlies the desynchronization between inner and outer time speed perception

In sum, both extremely low and high neuronal variability in SMN and VN lead to extreme manifestations of inner and outer time speed at the mental level: it is either too slow as in depression, or too fast as in mania. In contrast, average values in the SD of both SMN and VN yield a more or less balanced time speed perception – this allows for maximal synchronization of inner and outer time speed as in subjects without depression or mania. Accordingly, "average is good, extremes are bad" (Northoff and Tumati 2019).

The Self in Depression: Increased and Too Slow

Self and Cortical Midline Structures

Our sense of self is a central feature of our inner mental life. What do we mean by self? Psychological investigation shows a self-reference effect. Contents related to the self in various psychological functions, including memory, perception, emotions, and decision making, show higher accuracy and faster reaction in a behavioural task (Northoff 2016). In addition to self-reference, other cognitive functions related to the self include functions like self-experience, self-awareness, self-reflection, self-attribution, and self-monitoring – for the sake of simplicity, I will simply refer to sense of self, self-consciousness, or self-focus in a more or less synonymous way.

The sense of self is related to neuronal activity in specific regions of the brain. When asking participants to judge trait adjectives (own versus other) or other stimuli (like auto- versus heterobiographical events, or own versus other names), the resulting task-evoked activity strongly recruits the CMS (Northoff and Bermpohl 2004; Northoff et al. 2006). However, the CMS are not specific to the self, as they are implicated in other internal processes like emotion regulation, mind wandering, and social interaction.

Task-evoked activity in the CMS during self-reference strongly overlaps with resting-state activity in the DMN in specifically medial prefrontal cortex (including both ventral and dorsomedial prefrontal cortex; VMPFC and DMPFC) (Qin and Northoff 2011; Whitfiled-Gabrieli et al. 2011). This has been introduced as "rest-self overlap" (Bai et al. 2015) to describe the convergence between the self and the brain's spontaneous activity. Recent studies show that the spontaneous activity's temporal structure (as measured by PLE, autocorrelation window, and cross-frequency coupling) in the medial prefrontal cortex (MPFC) predicts the degree of self-consciousness (Huang et al. 2016; Wolff et al. 2019) (see chapter 4).

Depression: Increased Self and Its Prolonged Duration

Abnormal changes in our sense of self occur in, for instance, depression as in either major depressive disorder or as in the depressive phases of bipolar disorder. In these cases, the sense of self is abnormally increased and all thoughts revolve around it, which is associated with abnormally negative emotions like sadness and guilt. This has been described as "increased self-focus" or "increased internal focus" as a typical psychopathological hallmark of depression in general (Northoff and Sibille 2014). In contrast, the focus towards the external environment, described as "external environment-focus," is strongly reduced (Northoff and Sibille

2014). There is thus disbalance between increased self-focus and decreased environment-focus.

Various studies demonstrate that the CMS show atypically increased activity in depression during both resting state and tasks involving self-reference (Northoff 2016; Scalabrini et al. 2020). The neural activity is abnormally shifted: all activity is focused and increased in the CMS at the expense of the unimodal sensory regions. Nothing for free: if there is an increase somewhere in the brain, there needs to be a decrease somewhere else to compensate. The CMS thus acts like an abnormally strong magnet for the rest of the brain, including the sensory regions. For that reason, Andrea Scalabrini, a former student of mine and now a professor in Bergamo, titled his paper "All Roads Lead to the Default-Mode Network" (Scalabrini et al. 2020).

How can the CMS as part of the DMN act as super-strong magnet for the rest of the brain in depression? That is possible through its extremely strong slow frequencies – the slowest throughout the brain. In the depressed brain, they are even slower than in the non-depressed brain. From a dynamic point of view, we know that slower frequencies are more powerful: the increased slowness of the CMS/DMN in depression thus exerts an atypically strong power for the rest of the brain.

That atypical strong power also surfaces on the psychological or mental level. The self, mediated by the CMS, also exhibits atypically strong power in depression, the increased self-focus, whose abnormally long duration overshadows all other brain functions. Accordingly, we see yet another instance of how temporal features on the neuronal level translate onto the mental level: the slowness and high power in the CMS/DMN in depression resurfaces on the mental level in the increased self-focus and its prolonged duration.

Conclusion

Time has various facets. One of the most prominent facets of time is speed. Speed is not speed, however. Speed has different aspects. There is velocity. Our brain constructs velocity by, for instance, the neuronal variability of its amplitude. If that neuronal variability and thus the amplitude of neural activity is not changing at all, velocity and thus time speed is slow – we then experience time as extremely slow with nothing moving. That leads to depression in humans, where one experiences everything as too slow, which makes one unable to act at all. Depression can therefore be compared to a horse who remains unable to speed up at the start, falls behind early, or cannot leave the starting line at all.

Yet another aspect of time speed is duration. Duration describes the elapsed time; if that elapsed time is constructed by itself in an abnormally long way, any subsequent speed will be abnormally slow as it must cover that abnormally long temporal distance. That is the case in the self of depression. Here the brain constructs too-long durations – the neuronal moments of elapsed time are simply too long. The awareness of the own self is subsequently abnormally increased – an increased self-focus around which the thoughts and emotions revolve.

Together, we can see that time speed itself is not homogeneous. It includes different facets like velocity and duration. Each of these aspects seems to be mediated by different features and measures in the brain – changes in the latter can then lead to different changes in the consciousness of inner time speed. Most importantly, I have shown that the changes in the brain's construction of the different features of its time speed lead to different changes in inner time speed perception like in depression and mania. Even worse, these changes in time speed consciousness radically change the subjects' interaction with others and their respective environment. Time and more specifically time speed can thus be seen as the bridge or common currency between brain, consciousness, and world. I will elaborate in the next chapter.

6
Beyond Human Time

Introduction

We've come a long way. In the first chapter, I showed that the brain exhibits an inner time which itself constructs as distinguished from the world's outer time. I then demonstrated how the brain's inner time is key for consciousness in its relation to both the world (chapter 2) and the body (chapter 3). Reaching beyond consciousness to other mental features, I demonstrated that even our self is shaped by the brain's inner time, especially its longer timescales and slower frequencies (chapter 4). Finally, the brain's timescales are altered in mood disorders like depression and mania: they are either abnormally slow or fast, which results in the extremes of sadness and happiness (chapter 5).

Despite showing how the brain's inner time shapes our mental features, I have not yet tackled the elephant in the room: from where, and how, does the mind come and originate? Centuries of philosophers and neuroscientists have speculated about the mind and its relation to body and brain – the mind-body problem. Nowadays, we continue to speculate without a definite solution in sight.

My goal in this chapter is to provide a novel approach to this seemingly intractable and unsolvable problem. For that, I venture into the time of other non-human species as well as how time is key for creating artificial agents with mental features like a sense of self. This shows how timescales constitute an interface of world and brain/agent that is key for developing mental features. Based on these different lines of evidence, I propose that we no longer need the mind-body problem: it simply becomes superfluous, like pre-Darwin views of a God-based evolution once we discover biological nature and principles of evolution. We can then replace the mind-body problem with the world-brain problem that focuses on the temporal nature of their relationship and how that constitutes mental features.

Timescales in Other Species

Human and Non-human Species: Cross-Evolutionary Timescale Sharing

How about timescales in other species? Other non-human species share the same environment as we do. Moreover, other species like cats and dogs who live closely with humans are often our pets. There is lots of communication among the different species, including humans, most often in a non-verbal way. This raises the question of whether some sharing in the timescales among different species has occurred throughout evolution – and there is indeed some empirical evidence that it has.

The Japanese neuroscientist Shinomoto et al. (2009) investigated the firing rates, i.e., spike pattern, in different regions of the cortex in different species like cats, mice, rats, non-human primates (like monkeys), and humans. Interestingly, he found a very regular firing pattern in the motor cortex, a more random pattern in the visual cortex, and a burst-like pattern in the prefrontal cortex across all species. This was further confirmed by the fact that the differences in firing pattern were larger between regions across species than the ones between species within one region. Hence, different species share their firing patterns in their analogous regions. There seems to be some cross-evolutionary preservation of timescales shared by human and non-human species.

Such preservation holds on the cellular level, but also on the more systemic regional level of the whole brain. The neuroscientist Gyorgy Buszaki, together with colleagues including Nikos Logothetis and Wolf Singer (Buzsáki et al. 2013), demonstrates that different temporal features of neural activity like alpha rhythm (8 to 13 Hz), spindles, and ripples are present in different species including humans, non-human primates, dogs, bats, gerbils, guinea pigs, rabbits, mice, and hamsters. Moreover, these different species all show the same frequency range of these rhythmic patterns. They conclude: "In summary, the preservation of temporal constants that govern brain operations across several orders of magnitude of timescales suggests that the brain's architectural aspects – scaling of the ratios of neuron types, modular growth, system size, inter-system connectivity, synaptic path lengths, and axon caliber – are subordinated to a temporal organizational priority" (Buzsáki et al. 2013, 755).

Timescale Differences between Human and Non-human Species

There are similarities in timescales between humans and non-humans. But we as humans are not the same as monkeys let alone mice and rats. From where, then, do the differences come? The differences between species largely concern the ability to navigate in the environment. Bats can perceive ultrasonic waves and can thus navigate in certain environments, whereas humans lack the timescales for processing ultrasonic waves and cannot do that.

The environment presents us with a rich and extremely large repertoire of timescales, ranging from extremely slow ones, like seismic earth waves lasting decades, to extremely fast ones, like ultrasonic waves in bats lasting microseconds (Nagel 1974). The more timescales an organism has, the better it can align with the timescales of its environmental context. The organism may also use some tricks to amplify the efficacy of its limited number of timescales. It can connect different timescales with each other, which then may yield additional longer or shorter timescales in a dynamic repertoire (Golesorkhi et al. 2021a). Due to its extensive functional connectivity among the different regions and their timescales, the human brain seems to be particularly good in extending a dynamic repertoire beyond its actual number of timescales.

The organisms connect to the environment through their timescales. The more timescales with a larger dynamic repertoire an organism exhibits, the better the connection and alignment of that organism to the very large timescale repertoire of its environment. The better the organism can temporally align with its environment's timescales, the less errors it will make in navigating them, increasing the chance of its survival. Due to their sharing of timescales among each other, the different species share their survival in one and the same environment, the natural biological world. However, due to their differences in timescales, they create themselves and thus survive in different ecological niches of that shared environment. Finally, larger dynamic repertoire in timescales increases not only the chance of survival, but also dominance over other species with a more limited dynamic repertoire.

Timescales in Artificial Agents

Artificial Agents: Modularity and Representation

How about timescales in artificial agents (AA)? To understand the importance of timescales in AA, we first need to shed a brief light on the inner organization of AA and how it compares to that of the brain.

The traditional model of AA is module-based. Major AA proponents like Tony Prescott in the United Kingdom propose a modular model (Prescott and Camilleri 2019). The inner organization of the AA is composed of different modules: there is one module for movement, one for visual perception, one for executive functioning, one for recognizing other minds, one for memory, and so forth. Building these different modules within the agent's inner organization will then yield an AA which is highly functioning. Prescott and his group even assume that their modularity-based AA have a self, a humanoid self.

How is such module-based AA related to its environmental context? Here, the current forward-oriented artificial intelligence (AI) apostles look back towards a traditional philosophical model. Like past philosophers (and many current neuroscientists), they assume that the agents "represent" the external world. The agent is supposed to recapitulate the external world in its inner activity: the outer event that occurs outside, like the laptop falling out a window from the twenty-fifth floor, is re-created within the agents' inner activity and its pattern in a more or less one-to-one way, it is re-presented in the agent. This is also the predominant view of the brain in current neuroscience, which is often taken to represent its external environment in its inner neuronal activity.

Representation and modularity converge. Different components of the external environment are supposed to be represented in the different modules of the agent's inner organization. The action is represented in the movement module, the other's mind is represented in the module for recognizing other minds, the visual input is represented in the visual module, the flow of time from past to future is represented in the memory module, and so forth. Hence, modules represent distinct components of the environment.

The AI apostles are not alone. Many neuroscientists and psychologists as well as philosophers in our times subscribe, either implicitly or explicitly, to such view of the brain in terms of modularity and representation. The brain is composed of modules that serve to represent the world. There seems to be a closed loop between AI and neuroscience.

Brains: Topography of Different Timescales along a Uni- and Transmodal Gradient

However, the view of the brain I have sketched throughout this book looks rather different. Instead of modules, I have shown the importance of whole-brain organization. For instance, I did not localize consciousness in specific regions like anterior or posterior regions as often assumed by the current theories of consciousness like integrated information theory (IIT) and global neuronal workspace theory (GNWT) (Northoff and Lamme 2020). Instead, I assumed the whole brain with all of its regions to be involved in consciousness. Rather than single regions, the temporo-spatial theory of consciousness (TTC) assumes that the relationship and organization among all regions is key for consciousness – the more they are nested within each other, i.e., the higher the temporo-spatial nestedness, the better for our consciousness.

The assumption of separate modules within the brain's inner organization is here replaced by a topography of the whole brain. Applied to geography and planetary science, topography refers to the shape of the land. Applied to the brain, topography refers to the organization of the various regions across the whole cortex, the brain's surface – the shape of the brain. As it concerns the whole cortex with all its regions, topography cannot be chopped up into different modules – topography replaces modularity.

What does the topographic organization of the brain look like? One key feature is the continuum or transition between unimodal regions, like the visual and auditory cortex, and transmodal regions, like the ones of the default-mode network (DMN) – a uni-transmodal gradient. Such uni-transmodal topography converges with its temporal organization: unimodal regions show shorter timescales, or, shorter autocorrelation windows (ACW), than transmodal regions (Golesorkhi et al. 2021a; Golesorkhi et al. 2021b; Wolff et al. 2022). The uni-transmodal topography strongly impacts input processing: shorter timescales in unimodal sensory regions allow for higher temporal segregation of inputs, while the longer timescales in transmodal associative regions like the prefrontal cortex and the DMN are ideal for temporal integration (Wolff et al. 2022) (see chapter 1).

Let's take the example of music. Music includes shorter events, like special harmonies or brief rhythms that stand out. On the other hand, there are also melodies that span across longer timescales. Even longer timescales are present in the form of a sonata, where the initial melody may reoccur after five to twenty minutes towards the end of the movement. Our brain processes and encodes the different timescales of the music according to its own timescales: shorter events within the music are processed and temporally segregated by the shorter timescales of the brain's unimodal sensory cortex, while the longer events of melodies are encoded and temporally integrated by the longer timescales of brain's associative regions like the DMN (Wolff et al. 2022; Hasson et al. 2015).

Stochastic Matching of Timescales Replaces Representation

Do the brain's inner timescales re-present the outer event? No, there is no re-presentation or re-creation of the music' timescales by the brain's inner timescales. "An agent does not have a model of its world – it is a model. In other words, the form, structure, and states of our embodied brains do not contain a model of the sensorium – they are that model" (Friston 2013, 212).

Instead, there is matching between the inner and outer timescales: according to their statistical or stochastic distribution, inner and outer timescales match. If the music contains timescales that overlap and match with those of the brain, the latter can process and encode the former and enjoy the pleasure of music. If, in contrast, there is no stochastic overlap between the music's and the brain's timescales, the brain cannot process the music nor enjoy its pleasure. Stochastic matching of the whole brain's timescales with the music's timescales replaces module-based representation of the latter by the former.

How about stochastic matching in the case of AA? In order for such stochastic matching to be possible, we will need to build in different timescales into the inner organization or topography of AA. Learning from the brain, we can build in different timescales along the same gradient that is characterized by the gradient of uni-transmodal regions in the brain. Note that we do not need to build the unimodal sensory and transmodal associative regions themselves into our AA. We only need to build in the uni- and transmodal regions' timescales. That, I hypothesize, should be sufficient for the

AA to stochastically match its own inner timescales with those of the outer environment.

Now comes the litmus test. We can play music to the AA with the timescales that we as humans can match with and apprehend. If we build in only the slower timescales (those of the transmodal associative regions in the case of the human brain), our agent will not be able to stochastically match the music's faster timescales and will dance too slowly. If, in contrast, we only build in the faster timescales (those of the unimodal sensory regions), our agent will not be able to match the music's slower timescales and will dance too quickly. Hence, the agent requires the whole dynamic repertoire of timescales of the music to properly match and process it and consequently to dance to it.

How can we build in such timescale-based AA? We can build in different timescales within each processing unit. Even more exciting, we can even go beyond the timescales of humans. For instance, bats can process timescales related to ultrasonic waves. Bats consequently experience the environment in an ultrasonic way, very different from the way we as humans process it – "what it is like to be bat," as an index of their consciousness, is thus different from "what it is like to be human" (Nagel 1974). Building timescales into our AA that extend beyond those of our human brain will thus enable us to create a more capable agent, which may allow for more fine-grained and wider stochastic matching with the environmental context. One may even go so far and assume that the degree of temporal extension of the AA timescales beyond those of humans should go along with more or less similar degrees of extension in the capacities of such AA to operate within the same environment.

Do Timescale- and Matching-Based Artificial Agents Have an "Experience of Self"?

Recall that Prescott assumes that his module- and representation-based AA have a self (Prescott and Camilleri 2019). How about the here-proposed timescale- and matching-based agents? To address this question, we need to make a key distinction, namely between "having a self" and "experiencing a self." Having a self means that one has all the capacity of a self, which can include the ability to recognize other minds, autobiographical memories, and voluntarily initiate movements and action. Taken in this sense, Prescott's AA may indeed be attributed a self, that is, in the sense of "having a self."

However, the self of his robots may not be experienced as such by the robots themselves. They may not have a sense of self as an "experience of self," and may act like a self but may not be able to experience their own self as such. There is no consciousness of self, as to assume that would be to confuse "having a self" and "experiencing a self," which amounts to the difference between of self and sense of self. I claim that we require timescale- and matching-based AA rather than module- and representation-based AA to endow our AA with a sense of self, an experience of self.

Why do we need timescales and stochastic matching to have an experience or sense of self and, more generally, to have consciousness of both self and world? We experience our self as part of the wider world; we have a point of view within the world from which we experience ourselves as part of the world (Northoff and Smith 2022). Based on various lines of empirical evidence, we assume that such a point of view is constituted by the stochastic matching of the agent's timescales with those of the world. The better they match, the closer the point of view aligns our experience with the world itself; the more discrepant they are, the lower degree of alignment.

Since Prescott's module- and representation-based agents do not contain any inner timescales, stochastically they cannot match with and are not anchored within the environment. For that reason, I argue, they do not exhibit a point of view at all. Since they do not have a point of view, they cannot experience the world nor their own self as part of the world as whole, and because there is no experience of self, there is no sense of self.

How can we test that? One key feature of timescales is their fluctuations. I have shown that the brain's inner time and thus its timescales fluctuate over time and are highly variable. The same applies to mental features like our thoughts and even our self, which fluctuate over time in slower (Rostami et al. 2022) and faster (Hua et al. 2022) ways. Does the self of Prescott's agent fluctuate over time? I would assume that it does not fluctuate at all, as there are no timescales built in to his agents that could allow for such fluctuation to occur. The fluctuations of the different timescales make possible their stochastic matching with the environmental context and ultimately the constitution of a point of view. Hence, if our AA's experience of a self fluctuates over time (and is modulated by the fluctuations of the environment), we may have some empirical support for our assumption of a sense of self in our AA.

From the Mind-Body Problem to the World-Brain Problem

Timescales of the World, the Brain, and the Mind

I sketched a primarily temporal relationship of world and brain/agent. Both share some timescales across different species, and differ in others in a species-specific way. The temporal relationship of world and organism is thus characterized by both trans-species similarities and cross-species differences. Timescales are shared by world and brain/agent – they are their common currency (Northoff et al. 2020a; Northoff et al. 2020b).

This might be confusing. I introduced the concept of "common currency" to describe the shared features of the brain and the mind: both neural and mental features show similar timescales and dynamics. This is different from the shared features of the world and the brain/agent. However, conceived in a purely logical way, the fact that (i) brain and mind, as well as (ii) world and brain share their respective timescales entails that world and mind should also share their timescales. The world's timescales should be manifest, at least in part, in the timescales of our mind as mediated by the brain/agent. The timescales and their dynamics may be for world and the mind the common currency as mediated by the temporal relationship of the world and the brain.

There is plenty of empirical evidence to indicate that the timescales of the world shape the timescales of the mind. Once again consider music. We dance to the rhythm of the music, and our emotions fluctuate with its timescales: if they become slow we may feel sadness, while if they become fast we may feel excitement and joy. The world's timescales, in this case those of the music, resurface in our mind, its cognition, its emotions, its movements, etc. – this, I propose, is possible through the brain's timescales that mediate between world and mind.

Time Shapes the Mind: Common Currency of the World, the Brain, and the Mind

The outer timescales of the world fluctuate continuously, especially at faster time speeds. If these timescales shape our inner thoughts, then we would expect the latter to fluctuate, too. That is indeed the case, as two recent studies show.

Jingyu Hua (Hua et al. 2022) investigates thoughts related to an external task to which the subject had to pay attention (on-thoughts), and thoughts unrelated to the task as subjects did not pay attention (off-thoughts). Using EEG, he showed that on-thoughts related to fast-paced tasks were mediated by a faster frequency, alpha (8 to 13 Hz), compared to off-thoughts which were mediated by a slower frequency, theta (5 to 8 Hz). Hence, the brain uses faster frequencies and shorter timescales to mediate the faster input from the environment to shape its corresponding outer on-thoughts, and relies on slower frequencies to revert and detach itself in its off-thoughts.

Yet another purely psychological study by Rostami et al. (2022) shows the timescales of our thoughts. They asked subjects to indicate the switch between internal-oriented thoughts, such as about the self, and external-oriented thoughts, such as about the environment. Subjects shifted their thought around every twenty seconds, which indicates fluctuations and dynamic thought. Internally oriented thoughts also last longer than externally oriented thoughts, which operated on shorter timescales. This is complemented by different regions in the brain: externally oriented thoughts use the faster unimodal sensory regions, while internally oriented thoughts recruit the slower transmodal regions like the prefrontal cortex and the DMN (Vanhaudenhuyse et al. 2011).

Together, these studies demonstrate that timescales shape our thoughts, including their relationship to the brain and the world. Albeit tentatively, they lend support to the assumption that the world, the brain, and the mind share timescales and ultimately dynamic as their common currency. If such temporal sharing ceases, we lose consciousness, including our sense of self, as happens in deep sleep, anesthesia, or coma. If the degree of shared time among the world, the brain, and the mind shifts to longer timescales and slower frequencies, we become depressed. If it shifts to shorter timescales and faster frequencies, we become manic. Time thus shapes and constitutes both our mind and its relationship to the world.

From the Mind-Body Problem to the World-Brain Problem

We are now ready to tackle one of the most obstinate problems in philosophy and neuroscience: the mind-body problem. Centuries of philosophers and neuroscientists have suggested how mind and body are related. Descartes suggested dualism of mind and body, which was replaced by assuming that the mind is traced to the brain as part of the body. But the core of the problem remains nevertheless. Rather than searching for a special mind as distinct from the body, we now search for a special neuronal feature within the brain as distinct from the rest of the brain (see chapter 2). The specialness of the mind is thus preserved, and therefore so is the mind-body/brain problem.

However, the data in this and the previous chapters tell a different story. There is nothing special about the mind. Like the rest of the biological or natural world, the mind is based on time, namely the different timescales and their dynamic. The mind is a temporal phenomenon based on how well the brain aligns its inner time to the outer time of the world. The mind can thus be traced to the temporal nature of the relation of the world and the brain, the world-brain relation (Northoff 2018). This carries an important implication.

I claim that the temporal features of the world-brain relation, like their degree of stochastic matching, can explain the occurrence of mental features like consciousness and sense of self. The question of the mind-body problem then becomes obsolete and nonsensical: why question something, in this case the relationship of mind and body, that can already be explained by something more basic and fundamental, in this case the temporal relationship of world and brain?

Investigating the temporal features of the relationship of the world and the brain will tell us about the latter's mind. All we need to do is to investigate the timescales with their stochastic matching of world and brain. To return to the bat, if we know the timescales of ultrasonic waves, we know the timescales of the bat and that, in turn, tells us about the bat's mental features, or "what it is like to be a bat" (Nagel 1974). Nothing beyond timescales like an additional feature such as a separate mind or a special brain mechanism (like information integration or global neuronal workspace) is needed to explain mental features like consciousness, self, etc. The mind-body problem can thus be replaced by the world-brain problem (Northoff 2018).

Conclusion

I have shown that different species, human and non-human, share certain timescales and differ in others. Timescales shape our relationship to the world, and our behavioural and mental features – even mental features like self, consciousness, emotions, and others. Our relationship to the world is essentially temporal and this, in turn, shapes our mind. This carries major philosophical reverberations.

Once we know that time provides the basis for mental features through the timescales of world-brain relation, we no longer need the assumption of mind. Any assumption of mind will then look strange and nonsensical, in the same way that the historical assumption of an "elan vital," or life energy, before the discovery of the DNA, or the assumption of life spirits before the discovery of the pumping nature of the heart, now look nonsensical. That renders any subsequent question for the mind-body problem nonsensical: if something (the mind) is absent, it cannot stand in any kind of relationship to something (the body) that is present. Goodbye mind-body problem. We can replace the nonsensical mind-body problem with a more sensical one: the world-brain problem. Hello world-brain problem.

The world-brain problem describes the kind of relationship that is required for constituting mental features. I have demonstrated that such a relationship must be temporal and characterized by transforming the timescales of the world into the timescales of the mind through the timescales of the brain. Note that such a temporal transformative relationship is not limited to the brain. We could also establish the temporal basis of world-artificial agent relationship which, unlike the evolutionarily well-developed world-brain relation, remains to be realized.

Coda: Copernican Revolution in Neuroscience and Philosophy

World is time, brain is time, mind is time. Time is dynamic as it consists in the continuous construction of inner time. Inner time is about the temporal relationships between events or objects within both the world and the brain, as well as between the world and the brain themselves. Our brain's construction of its own inner time is part of the ongoing world's construction of its own inner time. Hence, by constructing its own inner time, the brain aligns its inner time to the world's inner time.

The brain integrates its inner time within and aligns its inner time to the world's inner time. This makes possible mental features like self and consciousness. If the brain's construction of its own inner time, including its alignment with the world, is altered or lost, our mental features like consciousness will also be altered or lost. This puts the brain's inner time in the midst of the change and dynamic passages of the world's inner time. Becoming part of the world as whole, the brain itself becomes temporal.

How can we view the brain as part of the world's time? We need to learn from Copernicus and Darwin. We used to conceive of ourselves as the centre. Before Copernicus, we conceived of the earth, as our location, as the centre of the universe. Copernicus, however, taught us otherwise through, as I say, thinking the earth inside and as part of the ongoing time of the universe. The earth is just one part of the solar system in which the sun, rather than earth, is the centre. The Copernican revolution is the shift that allowed us to understand why and how the earth moves. We could then understand how the earth and its movements in time are part of the universe as whole, featured by its ongoing passage of time.

The same applies analogously to Darwin. Before Darwin, we thought that the human species was the centre of biology and of other species – we, as humans, conceived of ourselves as special, atemporal, and God-given and thus eternal when compared to all other species. Darwin was intelligent enough to rob us of that illusion when thinking human species inside and as part of the ongoing passage of the evolution's time. We as humans are but one species among the many to evolve throughout the natural course of life and evolution. Darwin thus shifted the human species into the ongoing passage of time of biological evolution, making it temporal and thereby part of the ongoing passage of time of evolution.

Despite occurring in different fields, the Copernican revolutions in physics and biology share a commonality. We could understand the changes of both the earth and the human species much better once we conceived them in a temporal way. The movements of the earth are just one temporal part of the dynamic passage of time of the wider universe. The same holds for the human species. The human species is just one temporal part of the wider temporally more expanded evolution of all biological species as whole. The Copernican revolution in both physics and biology is thus an essentially temporal revolution, that is, how earth and human (and non-human) species became temporalized and thereby integrated within the wider world and its time (Weinert 2013; Northoff 2018; Northoff et al. 2019).

I propose that we require an analogous shift in our view of the brain and the mind. Neither is atemporal. I have shown how the brain constructs its own inner time and how that yields mental features like consciousness, self, and others. Moreover, the data show that the coupling or alignment of the brain's inner time to the world's inner time is key for yielding mental features. By becoming part of the more comprehensive and wider world's time, the brain's inner time can yield mental features. Accordingly, we can only understand the mind though the brain's alignment with the continuous passage of world time. Brain time is part of world time and that, in turn, makes possible the mind.

We need nothing less than a Copernican revolution in neuroscience and philosophy (Northoff 2018; 2019). We must abandon our pre-Copernican stance of conceiving brain-mind as the atemporal special centre of the world. Copernicus and Darwin would probably smile if they listened in when we talk about brain and mind. They might advise us to conceive of the brain's inner time and how it integrates with the continuous passage of world time. Time is of the essence, not only for earth and evolution but also for brain-mind.

In conclusion, we require a Copernican revolution in neuroscience and philosophy. Analogous to the revolutions in physics and biology by Copernicus and Darwin, the proclaimed Copernican revolution is essentially temporal — we need to conceive of the inner time of brain-mind as inside and part of the ongoing passage of the world's time. Once we understand the interface of the brain's inner time with world time, we can replace the mind-body problem with the world-brain problem. This renders superfluous the assumption of the mind as a special entity in the same way the Copernican revolutions in both physics and biology rendered spurious the assumption of God as a primary mover of the earth and a creator of humans.

References

Baars, Bernard J. 2005. "Global Workspace Theory of Consciousness: Toward a Cognitive Neuroscience of Human Experience." *Progress Brain Research* 150: 45–53. https://doi.org/10.1016/S0079-6123(05)50004-9.

Babo-Rebelo, Mariana, Craig G. Richter, and Catherine Tallon-Baudry. 2016. "Neural Responses to Heartbeats in the Default Network Encode the Self in Spontaneous Thoughts." *Journal of Neuroscience* 36, no. 30 (27 July): 7,829–40. https://doi.org/10.1523/JNEUROSCI.0262-16.2016.

Bai, Y., T. Nakao, J. Xu, P. Qin, P. Chaves, A. Heinzel, and G. Northoff. 2015. "Resting State Glutamate Predicts Elevated Pre-stimulus Alpha during Self-relatedness: A Combined EEG-MRS Study on 'Rest-Self Overlap.'" *Social Neuroscience* 11, no. 3. https://doi.org/10.1080/17470919.2015.1072582.

Berger, Hans. 1929. "About the Electroencephalogram of Huamns/Über das Elektrenkephalogramm des Menschen." *Archiv für Psychiatrie und Nervenkrankheiten* 87: 527–70.

Bergson, Henri. 1946. *The Creative Mind: An Introduction to Metaphysics.* (La Pensée et le mouvant, 1934). Citadel Press.

Bishop, Geo. 1933. "Cyclic Changes in Excitability of the Optic Pathway of the Rabbit." *American Journal of Physiology* 103: 213–24.

Buzsáki, Georgy, Nikos Logothetis, and Wolf Singer. 2013. "Scaling Brain Size, Keeping Timing: Evolutionary Preservation of Brain Rhythms." *Neuron* 80, 751–64.

Cairns, Hugh. 1941. "Head Injuries in Motor-Cyclists: The Importance of the Crash Helmet." *British Medical Journal* 2, no. 4213 (4 Oct.): 465–71.

Churchland, Patricia S. 2002. *Brain-Wise: Studies in Neurophilosophy.* Cambridge, MA: MIT Press.

Dainton, Barry. 2010. *Time and Space.* Montreal: McGill-Queen's University Press.

Damiani, Stefano, Andrea Scalabrini, Javier Gomez-Pilar, Natascia Brondino, and Georg Northoff. 2019. "Increased Scale-Free Dynamics in Salience Network in Adult High-Functioning Autism." *Neuroimage Clinical* 21, 101634. https://doi.org/10.1016/j.nicl.2018.101634.

Deco, Gustavo, Viktor Jirsa, A.R. McIntosh, Olaf Sporns, and Rolf Kötter. 2009. "Key Role of Coupling, Delay, and Noise in Resting Brain Fluctuations." *Proceedings of the National Academy of Sciences* 106, no. 25 (23 Jun.): 10302–7. https://doi.org/10.1073/pnas.0901831106.

Dehaene, Stanislas, and Jean-Pierre Changeux. 2011. "Experimental and Theoretical Approaches to Conscious Processing." *Neuron* 70, no. 2 (28 Apr.): 200–27. https://doi.org/10.1016/j.neuron.2011.03.018.

Dehaene, Stanislas, Lucie Charles, Jean-Rémi King, and Sébastien Marti. 2014. "Toward a Computational Theory of Conscious Processing." *Curr. Opin. Neurobiol.* 25: 76–84. https://doi.org/10.1016/j.conb.2013.12.005.

Dehaene, Stanislas, Hakwan Lau, and Sid Kouider. 2017. "What Is Consciousnes, and Could Machines Have It?" *Science* 358, no. 6362 (27 Oct.): 486–92. https://doi.org/10.1126/science.aan8871.

De Pasquale, Francesco, Stefania Della Penna, Abraham Z. Snyder, Laura Marzetti, Vittorio Pizzella, Gian Luca Romani, and Maurizio Corbetta. 2012. "A Cortical Core for Dynamic Integration of Functional Networks in the Resting Human Brain." *Neuron* 74, no. 4: 753–64.

Edelman, Gerald M., Joseph A. Gally, and Bernard J. Baars. 2011. "Biology of Consciousness." *Frontiers in Psychology* 2, no. 4 (Jan. 25). https://doi.org/10.3389/fpsyg.2011.00004.

Fingelkurts, A.A., A.A. Fingelkurts, and C.F.H. Neves. 2010 "Natural World Physical, Brain Operational, and Mind Phenomenal Space-Time." *Phys. Life Rev.* 7, no. 2: 195–249. https://doi.org/10.1016/j.plrev.2010.04.001.

Fingelkurts, Andrew A., Alexander A. Fingelkurts, Sergio Bagnato, Cristina Boccagni, and Giuseppe Galardi. 2013. "Dissociation of Vegetative and Minimally Conscious Patients Based on Brain Operational Architectonics: Factor of Etiology." *Clinical EEG and Neuroscience* 44, no. 3: 209–20. https://doi.org/10.1177/1550059412474929.

Friston, Karl. 2010. "The Free-Energy Principle: A Unified Brain Theory?" *Nat Rev Neurosci* 11: 127–38. https://doi.org/10.1038/nrn2787.

– 2013. "Life As We Know It." *J.R. Soc. Interface* 10: 20130475. https://doi.org/10.1098/rsif.2013.0475.

Fuchs, Thomas. 2013. "Temporality and Psychopathology." *Phenomenology in the Cognitive Sciences* 12: 75–104.

Goldstein, Kurt. 2000. *The Organism: A Holistic Approach to Biology Derived from Pathological Data in Man.* Reprint. New York: Zone Books/MIT Press.

Golesorkhi, M., J. Gomez-Pilar, S. Tumati, M. Fraser, and G. Northoff. 2021a. "Temporal Hierarchy of Intrinsic Neural Timescales Converges with Spatial Core-Periphery Organization." *Commun. Biol.* 4, 277.

Golesorkhi, M., J. Gomez-Pilar, F. Zilio, N. Berberian, A. Wolff, M. Yagoub, and G. Northoff. 2021b. "The Brain and Its Time: Intrinsic Neural Timescales Are Key for Input Processing." *Commun. Biology* 4: 970. https://doi.org/10.1038/s42003-021-02483-6.

Hardstone, Richard, Simon-Shlomo Poil, Giuseppina Schiavone, Rick Jansen, Vadim V. Nikulin, Huibert D. Mansvelder, and Klaus Linkenkaer-Hansen. 2012. "Detrended Fluctuation Analysis: A Scale-Free View on Neuronal Oscillations." *Frontiers in Physiology* 3 (30 Nov.): 450. https://doi.org/10.3389/fphys.2012.00450.

Hasson, Uri, Janice Chen, and Christopher J. Honey. 2015. "Hierarchical Process Memory: Memory as an Integral Component of Information Processing." *Trends in Cognitive Sciences* 19, no. 6: 304–13.

He, Biyu J. 2014. "Scale-Free Brain Activity: Past, Present, and Future." *Trends in Cognitive Sciences* 18, no. 9 (Sept.): 480–7. https://doi.org/10.1016/j.tics.2014.04.003. Review.

He, Biyu J., John M. Zempel, Abraham Z. Snyder, and Marcus E. Raichle. 2010. "The Temporal Structures and Functional Significance of Scale-Free Brain Activity." *Neuron* 66, no. 3 (13 May): 353–69. https://doi.org/10.1016/j.neuron.2010.04.020.

Hipp, Joerg F., David J. Hawellek, Maurizio Corbetta, Markus Siegel, and Andreas K. Engel. 2012. "Large-Scale Cortical Correlation Structure of Spontaneous Oscillatory Activity." *Nature Neuroscience* 15, no. 6: 884–90. https://doi.org/10.1038/nn.3101.

Hua, J., A. Wolff, J. Zhang, L. Yao, Y. Zang, J. Luo, X. Ge, C. Liu, and G. Northoff. 2022. "Alpha and Theta Peak Frequency Track on- and off-Thoughts." *Communications Biology* 5, no. 1: 1–13. https://doi.org/10.1038/s42003-022-03146-w.

Huang, Zirui, Natsuho Obara, Henry Hap Davis 4th, Johanna Pokorny, and Georg Northoff. 2016. "The Temporal Structure of Resting-State Brain Activity in the Medial Prefrontal Cortex Predicts Self-Consciousness." *Neuropsychologia* 82 (Feb.): 161–70. https://doi.org/10.1016/j.neuropsychologia.2016.01.025.

James, William. 1890. *Principles of Psychology*. Cambridge, MA: Harvard University Press.

Kolvoort, Ivar, Soeren Wainio-Theberge, Annemarie Wolff, and Georg Northoff. 2020. "Temporal Integration as 'Common Currency' of Brain and Self-Scale-Free Activity in Resting-State EEG Correlates with Temporal

Delay Effects on Self-Relatedness." *Hum Brain Mapp.* 41: 4355–74.

Lakatos, Peter, James Gross, and Gregor Thut. 2019. "A New Unifying Account of the Roles of Neuronal Entrainment." *Curr. Biol.* 29: R890–R905.

Lamme, V.A.F. 2018. "Challenges for Theories of Consciousness: Seeing or Knowing, the Missing Ingredient and How to Deal with Panpsychism." *Philos. Trans. Biol. Sci.* 373: 20170344. https://doi.org/10.1098/rstb.2017.0344.

Lashley, K. 1951. "The Problem of Serial Order in Behavior." http://faculty.samford.edu/~sfdonald/Courses/cosc470/Papers/The%20problem%20of%20serial%20order%20in%20behavior%20(Lashley).pdf.

Lau, H., and D. Rosenthal. 2011. "Empirical Support for Higher-Order Theories of Conscious Awareness." *Trends Cogn. Sci.* 15: 365–73. https://doi.org/10.1016/j.tics.2011.05.009.

Lechinger, Julia, Dominik Philip Johannes Heib, Walter Gruber, Manuel Schabus, and Wolfgang Klimesch. 2015. "Heartbeat-Related EEG Amplitude and Phase Modulations from Wakefulness to Deep Sleep: Interactions with Sleep Spindles and Slow Oscillations." *Psychophysiology* 52, no. 11: 1441–50. https://doi.org/10.1111/psyp.12508.

Linkenkaer-Hansen, Klaus, Vadim V. Nikouline, J. Matias Palva, and Risto J. Ilmoniemi. 2001. "Long-Range Temporal Correlations and Scaling Behavior in Human Brain Oscillations." *Journal of Neuroscience* 21, no. 4: 1370–7.

McGinn, C. 1991. *The Problem of Consciousness.* London: Blackwell.

Mashour, G.A., P. Roelfsema, J.-P. Changeux, and S. Dehaene. 2020. "Conscious Processing and the Global Neuronal Workspace Hypothesis." *Neuron* 105: 776–98. https://doi.org/10.1016/j.neuron.2020.01.026.

Meisel, C., A. Klaus, V.V. Vyazovskiy, and D. Plenz. 2017. "The Interplay between Long- and Short-Range Temporal Correlations Shapes Cortex Dynamics across Vigilance States." *J. Neurosci.* 37, 10114–24. https://doi.org/10.1523/JNEUROSCI.0448-17.2017.

Metzinger, Thomas. 2003. *Being No One.* Cambridge, MA: MIT Press.

Monto, Simo. 2012. "Nested Synchrony: A Novel Cross-Scale Interaction among Neuronal Oscillations." *Frontiers in Physiology* 3.

Monto, Simo, Satu Palva, Juha Voipio, and J. Matias Palva. 2008. "Very Slow EEG Fluctuations Predict the Dynamics of Stimulus Detection and Oscillation Amplitudes in Humans." *Journal of Neuroscience* 28, no. 33: 8268–72. https://doi.org/10.1523/JNEUROSCI.1910-08.2008.

Murray, Ryan J., Marie Schaer, and Martin Debbané. 2012. "Degrees of Separation: A Quantitative Neuroimaging Meta-analysis Investigating Self-Specificity and Shared Neural Activation between Self- and Other-Reflection." *Neuroscience & Biobehavioral Reviews* 36, no. 3: 1043–59. https://doi.org/10.1016/j.neubiorev.2011.12.013.

Murray, Ryan J., Martin Debbané, Peter T. Fox, Danilo Bzdok, and Simon B. Eickhoff. 2015. "Functional Connectivity Mapping of Regions Associated with Self- and Other-Processing." *Human Brain Mapping* 36, no. 4: 1304–24. https://doi.org/10.1002/hbm.22703.

Nagel, Thomas. 1974. "What Is It Like to Be a Bat?" *Philosophical Review* 83, no. 4: 435–50.

Northoff, Georg. 2004. *Philosophy of Brain*. Amsterdam: John Benjamins.

– 2011. *Neuropsychoanalysis in Practice: Self, Objects, and Brains*. Oxford: Oxford University Press.

– 2012a. "Immanuel Kant's Mind and the Brain's Resting State." *Trends in Cognitive Science* 16, no. 7: 356–9. https://doi.org/10.1016/j.tics.2012.06.001.

– 2012b. *Was nun Herr Kant?* (What's up Mister Kant?) New York: Random House.

– 2013. "What the Brain's Intrinsic Activity Can Tell Us about Consciousness: A Tri-dimensional View." *Neuroscience & Biobehavioral Reviews* 37, no. 4: 726–38.

– 2014a. *Unlocking the Brain, Vol I: Coding*. Oxford: Oxford University Press.

– 2014b. *Unlocking the Brain, Vol II: Consciousness*. Oxford: Oxford University Press.

– 2015. "Do Cortical Midline Variability and Low Frequency Fluctuations Mediate William James' 'Stream of Consciousness'? 'Neurophenomenal Balance Hypothesis' of 'Inner Time Consciousness.'" *Consciousness and Cognition* 30: 184–200. https://doi.org/10.1016/j.concog.2014.09.004.

– 2016. *Neurophilosophy and the Healthy Mind: Learning from the Unwell Brain*. New York: Norton.

– 2018. *The Spontaneous Brain: From the Mind-Body to the World-Brain Problem*. Cambridge, MA: MIT Press.

– 2019. "Lessons from Astronomy and Biology for the Mind: Copernican Revolution in Neuroscience." *Frontiers in Human Neuroscience* 13: 319. https://doi.org/10.3389/fnhum.2019.00319.

Northoff, G., and F. Bermpohl. 2004. "Cortical Midline Structures and the Self." *Trends in Cognitive Sciences* 8, no. 3: 102–07. https://doi.org/10.1016/j.tics.2004.01.004.

Northoff, G., A. Heinzel, M. de Greck, F. Bermpohl, H. Dobrowolny, and J. Panksepp. 2006. Self-Referential Processing in Our Brain: A Meta-analysis of Imaging Studies on the Self." *Neuroimage* 31, no. 1: 440–57. https://doi.org/10.1016/j.neuroimage.2005.12.002.

Northoff, Georg, and Zirui Huang. 2017. "How Do the Brain's Time and Space Mediate Consciousness and Its Different Dimensions? Temporo-spatial Theory of Consciousness (TTC)." *Neuroscience & Biobehavioral Reviews* 80: 630–45. https://doi.org/10.1016/j.neubiorev.2017.07.013.

Northoff, G., and V. Lamme. 2020. "Neural Signs and Mechanisms of Consciousness: Is There a Potential Convergence of Theories of Consciousness in Sight." *Neurosci. Biobehav. Rev.* 118: 568–87. https://doi.org/10.1016/j.neubiorev.2020.07.019.

Northoff, Georg, Paola Magioncalda, Matteo Martino, Hsin-Chien Lee, Ying-Chi Tseng, and Timothy Lane. 2018. "Too Fast or Too Slow? Time and Neuronal Variability in Bipolar Disorder: A Combined Theoretical and Empirical Investigation." *Schizophrenia Bulletin* 44, no. 1: 54–64. https://doi.org/10.1093/schbul/sbx050.

Northoff, Georg, and David Smith. 2022. "The Subjectivity of Self and Its Ontology: From the World-Brain Relation to the Point of View in the World." *Theory & Psychology* 1–30. https://doi.org/10.1177/09593543221080120.

Northoff, Georg, and Etienne Sibille. 2014. "Why Are Cortical GABA Neurons Relevant to Internal Focus in Depression? Across-Level Model Linking Cellular, Biochemical and Neural Network Findings." *Mol. Psychiatry* 19: 966–77.

Northoff, G., and S. Tumati. 2019. "'Average Is Good, Extremes Are Bad': Non-linear Inverted U-shaped Relationship between Neural Mechanisms and Functionality of Mental Features." *Neurosci. Biobehav. Rev.* 104: 11–25. https://doi.org/10.1016/j.neubiorev.2019.06.030.

Northoff, G., D. Vatansever, A. Scalabrini, and E.A. Stamatakis. 2022. "Ongoing Brain Activity and Its Role in Cognition: Dual versus Baseline Models." *The Neuroscientist*, 1–28. https://doi.org/10.1177/10738584221081752.

Northoff, G., S. Wainio-Theberge, and K. Evers. 2020a. "Is Temporo-spatial Dynamics the 'Common Currency' of Brain and Mind? In Quest of 'Spatiotemporal Neuroscience.'" *Physics of Life Reviews* 1: 1–21. https://doi.org/10.1016/j.plrev.2019.05.002.

Northoff, G., S. Wainio-Theberge, and K. Evers. 2020b. "Spatiotemporal Neuroscience: What Is It and Why We Need It." *Physics of Life Reviews* 33: 78–87. https://doi.org/10.1016/j.plrev.2020.06.005.

Northoff, G., and F. Zilio. 2022a. "Temporo-spatial Theory of Consciousness (TTC): Bridging the Gap of Neuronal Activity and Phenomenal States." *Behavioral Brain Research* 424: 113788.

Northoff, G., and F. Zilio. 2022b. "From Shorter to Longer Timescales: Converging Integrated Information Theory (IIT) with the Temporo-spatial Theory of Consciousness (TTC)." *Entropy* 24, 270. https://doi.org/10.3390/e24020270.

Park, Hyeong-Dong, Stéphanie Correia, Antoine Ducorps, and Catherine Tallon-Baudry. 2014. "Spontaneous Fluctuations in Neural Responses to Heartbeats Predict Visual Detection." *Nature Neuroscience* 17, no. 4: 612–18. https://doi.org/10.1038/nn.3671.

Prescott, T.J., and D. Camilleri. 2019. "The Synthetic Psychology of the Self." *Cognitive Architectures* 94: 85–104.

Qin, P., and G. Northoff. 2011. "How Is Our Self Related to Midline Regions and the Default-Mode Network?" *Neuroimage* 57, no. 3: 1221–33. https://doi.org/10.1016/j.neuroimage.2011.05.028.

Qin, P., M. Wang, and G. Northoff. 2020. "Linking Bodily, Environmental and Mental States in the Self: A Three-Level Model Based on a Meta-analysis." *Neuroscience & Biobehavioral Reviews* 115: 77–95. https://doi.org/10.1016/j.neubiorev.2020.05.004.

Raichle, M.E., A.M. MacLeod, A.Z. Snyder, W.J. Powers, D.A. Gusnard, and G.L. Shulman. 2001. "A Default Mode of Brain Function." *Proceedings of the National Academy of Sciences of the United States of America*. https://doi.org/10.1073/pnas.98.2.676.

Raichle, Marcus E. 2009. "A Brief History of Human Brain Mapping." *Trends in Neurosciences* 32, no. 2: 118–26.

— 2015. "The Brain's Default Mode Network." *Annual Review of Neuroscience* 8, no. 38: 433–7. https://doi.org/10.1146/annurev-neuro-071013-014030. Review.

Richter, C.G., Mariana Babo-Rebelo, Denis Schwartz, and Catherine Tallon-Baudry. 2017. "Phase-Amplitude Coupling at the Organism Level: The Amplitude of Spontaneous Alpha Rhythm Fluctuations Varies with the Phase of the Infra-Slow Gastric Basal Rhythm." *Neuroimage* 146 (Feb. 1): 951–8. https://doi.org/10.1016/j.neuroimage.2016.08.043.

Rostami, S., A. Borjali, H. Eskandari, R. Rostami, A. Scalabrini, and G. Northoff. 2022. "Slow and Powerless Thought Dynamic Relates to Brooding in Unipolar and Bipolar Depression." *Psychopathology*, 1–15. https://doi.org/10.1159/000523944.

Rovelli, Carlos. 2018. *The Order of Time*. London: Penguin Random House.

Scalabrini, A., B. Vai, S. Poletti, S. Damiani, C. Mucci, C. Colombo, and M. Bendetti. 2020. "All Roads Lead to the Default-Mode Network: Global Source of DMN Abnormalities in Major Depressive Disorder." *Neuropsychopharmacology* 45, no. 12: 2058–69.

Seth, Anil, and Tim Bayne. 2022. "Theories of Consciousness." *Nature Reviews Neuroscience*. https://doi.org/10.1038/s41583-022-00587-4.

Shinomoto, S., et al. 2009. "Relating Neuronal Firing Patterns to Functional Differentiation of Cerebral Cortex." *PLoS Comput. Biol.* 5: e1000433.

Smith, David, Annemarie Wolff, Angelika Wolman, Julia Ignazsewski, and Georg Northoff. 2022. "Temporal Continuity of Self: Long Autocorrelation Windows Mediate Self-Specificity." *Neuroimage* 257, no 1: 119305.

Smolin, Lee. 2013. *Time Reborn: From the Crisis in Physics to the Future of the Universe*. Boston: Houghton Mifflin Harcourt.

Stevens, M.C., K.A. Kiehl, G. Pearlson, and V.D. Calhoun. 2007. "Functional Neural Circuits for Mental Timekeeping." *Hum Brain Mapp.* 28, no. 5: 394–408. https://doi.org/10.1002/hbm.20285.

Sui, Jie, and Glyn Humphreys. 2016. "Introduction to Special Issue: Social Attention in Mind and Brain." *Cogn Neurosci.* 7, no. 1–4: 1–4. https://doi.org/10.1080/17588928.2015.1112773.

Tagliazucchi, E., F. Von Wegner, A. Morzelewski, V. Brodbeck, K. Jahnke, and H. Laufs. 2013. "Breakdown of Long-Range Temporal Dependence in Default Mode and Attention Networks during Deep Sleep." *Proceedings of the National Academy of Sciences* 110, no. 38: 15419–24. https://doi.org/10.1073/pnas.1312848110.

Tagliazucchi, E., D.R. Chialvo, M. Siniatchkin, et al. 2016. "Large-Scale Signatures of Unconsciousness Are Consistent with a Departure from Critical Dynamics." *Journal of the Royal Society Interface* 13, no. 114: 20151027. https://doi.org/10.1098/rsif.2015.1027.

Tallon-Baudry, Catherine, Florence Campana, Hyeong-Dong Park, and Mariana Babo-Rebelo. 2018. "The Neural Monitoring of Visceral Inputs, rather than Attention, Accounts for First-Person Perspective in Conscious Vision." *Cortex* 102 (May): 139–49. https://doi.org/10.1016/j.cortex.2017.05.019.

Tipples, Jason, Victoria Brattan, and Pat Johnston. 2013. "Neural Bases for Individual Differences in the Subjective Experience of Short Durations (Less than 2 Seconds)." *PLoS One.* https://doi.org/10.1371/journal.pone.0054669.

Tononi, Giulio, Melanie Boly, Marcello Massimini, and Christof Koch. 2016. "Integrated Information Theory: From Consciousness to Its Physical Substrate." *Nature Reviews Neuroscience* 17, no. 7 (July): 450–61. https://doi.org/10.1038/nrn.2016.44.

Vanhaudenhuyse, Audrey, Athena Demertzi, Manuel Schabus, et al. 2011. "Two Distinct Neuronal Networks Mediate the Awareness of Environment and of Self." *Journal of Cognitive Neuroscience* 23, no. 3: 570–8. https://doi.org/10.1162/jocn.2010.21488.

Weinert, Friedel. 2013. *The March of Time.* Heidelberg: Springer.

Whitfield-Gabrieli, Susan, Joseph M. Moran, Alfonso Nieto-Castañón, Christina Triantafyllou, Rebecca Saxe, and John D.E. Gabrieli. 2011. "Associations and Dissociations between Default and Self-Reference Networks in the Human Brain." *Neuroimage* 55, no. 1: 225–32. https://doi.org/10.1016/j.neuroimage.2010.11.048.

Wolff, Annemarie, Daniel A. Di Giovanni, Javier Gómez-Pilar, Takashi Nakao, Zirui Huang, André Longtin, and Georg Northoff. 2019. "The Temporal Signature of Self: Temporal Measures of Resting-State EEG Predict Self-Consciousness." *Human Brain Mapping* 40, no. 3 (Feb. 15): 789–803. https://doi.org/10.1002/hbm.24412.

Wolff, Annemarie, Nareg Berberian, Mehrshad Golesorkhi, Javier Gomez-Pilar, Federico Zilio, and Georg Northoff. 2022. "Intrinsic Neural Timescales: Temporal Integration and Segregation." *Trends in Cognitive Sciences* 26, no. 2: 159–73. https://doi.org/10.1016/j.tics.2021.11.007.

Zhang, Jianfeng, Zirui Huang, and Yali Chen, et al. 2018. "Breakdown in the Temporal and Spatial Organisation of Spontaneous Brain Activity during General Anesthesia." *Human Brain Mapping* 39, no. 5: 2035–46. https://doi.org/10.1002/hbm.23984.

Zilio, Federico, Javier Gomez-Pilar, Shumei Cao, Jun Zhang, Di Zang, Zengxin Qi, Jiaxing Tan, Tanigawa Hiromi, Xuehai Wu, Stuart Fogel, Zirui Huang, Matthias R. Hohmann, Tatiana Fomina, Matthis Synofzik, Moritz Grosse-Wentrup, Adrian M. Owen, and Georg Northoff. 2021. "Are Intrinsic Neural Timescales Related to Sensory Processing? Evidence from Abnormal Behavioral States." *Neuroimage* 226. https://doi.org/10.1016/j.neuroimage.2020.117579.

Index

ancient Greeks and time, 3–7, 8, 9
anesthesia. *See* consciousness, loss or reduced
animals. *See* non-human species
artificial intelligence (AI) and artificial agents (AA), 10–11, 87, 101; artificial agents, 10–11, 87, 90–7, 99–101; brain topography, 91–4; dynamic repertoire in timescales, 89, 94; inner and outer perspective, 95–6; mental features, 10–11, 87, 90–4, 95–6; modules and representation, 90–6; music example, 92–4; philosophical models, 90–1; timescales in artificial agents, 90–7, 99–101
attention. *See* mental features
autocorrelation windows (ACW), 14, 24–5, 62, 66, 82, 92

bats, 52, 88, 89, 94, 100. *See also* non-human species
Berger, Hans, 22
Bergson, Henri, 5, 70–1, 73
biological evolution, 102–5
bipolar disorder, 80, 82. *See also* depression; mood disorders
Bishop, George H., 22
body, 42–7; brain's spontaneous activity, 45, 46; common currency of time, 46–7, 55–7, 97; consciousness, 36, 50, 55–7; cross-frequency coupling with organs, 45; double directionality of body and brain, 57; first- and third-person perspective, 36, 50, 52–3; inner time, 42–4, 57; LIS (locked-in syndrome), 37–40; mind-body and world-brain problem overview, 96–105; mind-brain-body dualism, 37, 40–2, 50, 52–4, 99–101; music example, 36–7, 97; neuroscience, 43–5, 61–2; stomach-brain relation, 42, 45–7; TTC (temporo-spatial theory of consciousness) overview, 35, 55–7; world-brain relation, 7–10, 11, 16–17, 36–7, 40–3, 44–7, 56–7. *See also* brain; mind-body and world-brain problem; motor functions; sensory functions; temporo-spatial theory of consciousness (TTC)
body, analogies: language, 46–7; surfing, 11, 41–3, 73; tango, 27, 43, 44, 57
body, heart: body inner time, 42, 43–7; brain's processing of heartbeat, 48–9, 52; common currency of time, 46–7, 49; consciousness, 48–9, 50, 52; HEP (heartbeat-evoked potential), 48–9, 50; neuro-cardiac coupling, 44, 45; tango analogy, 44. *See also* body

INDEX

brain, 12–25; active vs passive models, 20–5; brainwave-worldwave relation, 38, 42–3; common currency of time, 23–4, 25, 40–1, 46–7, 55–7, 96–105; double directionality of body and brain, 57; key questions, 7, 8, 9, 12; mind-body and world-brain problem overview, 96–105; mind-brain-body dualism, 37, 40–2, 50, 52–4, 99–101; self-similarity of world and brain, 31–3; TTC (temporo-spatial theory of consciousness) overview, 35, 55–7; world-brain relation, 7–10, 11, 14, 16–17, 24–5, 40–3. *See also* body; consciousness; mind; mind-body and world-brain problem; temporo-spatial theory of consciousness (TTC)

brain, analogies: horse racing, 74–6, 84; language, 46–7; ocean waves, 6–7, 9, 41–2, 63, 73; surfing, 11, 35, 41–3, 73; tango, 27, 42, 43, 57; walking, 33–5; water states, 8; windows, 14

brain, inner time, 6–7, 9–10, 13–17, 26–7, 38, 55–7, 102–5; accuracies/inaccuracies, 32–3; bodywaves, 57; brainwave-worldwave relation, 38, 42–3; change and persistence, 8–9, 13, 25, 30–1, 56; consciousness, 26–7, 33–5, 38, 57; correlated brain and world, 33–5, 38; as part of world's inner time, 9, 12–13, 24–5, 97, 102–5; self overview, 11, 60–9, 72–3; spontaneous activity, 14–15; synchronization of inner and outer time, 78, 80–1; temporal windows for world stimuli, 13–14, 19–20; world-brain relation, 7–10, 11, 13–14, 16–17, 27, 32–5, 38, 40–3, 56–7. *See also* body; duration of time; speed of time; spontaneous activity in brain; time, as common currency

brain, neuroscience, 17–25; active vs passive models of brain, 20–5; ACW (autocorrelation windows), 14, 24–5, 62, 66, 82, 92; CFC (cross-frequency coupling), 15, 45, 62, 82; convergence of temporal duration and spatial extension, 18–20; frequencies (slow-fast), 13–15, 26; LRTC (long-range temporal correlation), 17, 33–4; neuroimaging, 18, 28; PACC (perigenual anterior cingulate cortex), 60–5; PCC (posterior cingulate cortex), 60–5; scale-free activity, 14–17, 26, 31–4; spatial structure of spontaneous activity, 17–18; timescales, 14, 16, 26, 87–9, 96–8; timescale stochastic matching, 93–6; topography, 91–3; whole-brain organization, 18, 88, 91–3. *See also* consciousness, neuroscience; duration of time; nestedness; neuroscience; scale-free activity, neuroscience; self, neuroscience; speed of time; spontaneous activity in brain, neuroscience

Brown, Thomas Graham, 22, 23
Buszaki, Gyorgy, 88

cauliflower (Roman) analogy, 32
CFC (cross-frequency coupling), 15, 45, 62, 82
Chronos (continuum of time), 3–7
CMS (cortical midline structures), 60–3, 81–3
cognition; baseline model, 51; brain's spontaneous activity, 49, 50, 51; DMC (dual model of cognition), 51; DMN (default-mode network), 22–3; internal and external content, 51; LIS (locked-in syndrome), 37–40; neural correlates, 21; passive model of the brain, 20–1; sense of self, 81–2. *See also* consciousness
coma, 9, 26, 27, 28, 38–9, 99. *See also* consciousness, loss or reduced

common currency. *See* time, as common currency

communication and currency analogy, 46–7

consciousness, 7–10, 26–31, 35–41, 52–7; accuracies/inaccuracies, 32–3; active vs passive models of brain, 20–5; artificial agents (AA), 95–6; brain's inner time, 14, 16–17, 102–5; change and persistence, 30–1, 56–7, 58–60; clouded consciousness, 29, 30, 35; common currency of time, 30–1, 36–8, 40–1, 46–7, 55–7, 96–9; correlated brain and world, 16–17, 33–5, 38, 40–1, 56–7; dynamic time as basis for, 7–10, 25, 55–7, 102–5; first- and third-person perspective, 36, 50, 52–3, 70–1, 95–6; key questions, 7, 8, 9, 36, 49; LIS (locked-in syndrome), 37–40; mind-body and world-brain problem overview, 96–105; mind-brain-body dualism, 37–8, 50, 52–4, 99–101; non-human species, 10; philosophers on, 20, 50–4; self-similarity of world and brain, 31–3; special vs non-special mechanism, 54–7, 99, 104–5; TTC (temporo-spatial theory of consciousness), 35, 55–7, 91; wakefulness, 27–8; world-brain relation, 7–10, 11, 16–17, 27, 31–41, 43, 44, 55–7. *See also* mental features; mind-body and world-brain problem; temporo-spatial theory of consciousness (TTC)

consciousness, analogies: autopilot, 29; homogeneous soup, 28–9; language, 46–7; river or stream, 8, 9, 31, 77–8; surfing, 11, 35, 41–3, 73; tango, 27, 38, 42, 57; walking, 29, 33–5

consciousness, contents, 47–52; brain-heart relation, 48–9, 52; brain's spontaneous activity, 49–50, 51–2; external contents from world, 47–50, 57; internal contents from self, 47–51, 57; philosophers on, 50–4; tango analogy, 57; world-brain relation, 49–50, 57

consciousness, extended, 26–31, 36–7; boundary extensions, 30; common currency of time, 30, 40–1, 46–7; frequencies (slow-fast), 29–30; hyperawareness, 26–7; meditation, 33, 41; neuroscience, 36–7; psychostimulants, 27, 29, 34–5, 41; self-similarity of world and brain, 31–3; sleep deprivation, 27–8, 30

consciousness, loss or reduced, 26–31, 38–9, 98–9, 102; anesthesia, sleep, or coma, 9, 26–8, 30, 38–9, 98–9; autopilot analogy, 29; clouded consciousness, 29, 30, 35; common currency of time, 30, 40–1, 46–7, 98–9; frequencies (slow-fast), 28, 29, 30; homogeneous soup analogy, 28–9; loss of world-brain dynamics, 9, 26–9, 34, 36–8, 40–1, 98–9; scale-free activity disruptions, 17, 27–30, 32–3, 34; self-similarity of world and brain, 31–3; sensory and motor functions, 38–9; timescales, 28–9; URWS (unresponsive wakefulness), 39. *See also* sleep

consciousness, neuroscience, 26–31, 37–8; DFA (detrended fluctuation analysis), 27–30; frequencies (slow-fast), 13–14, 28; GNWT (global neuronal workspace theory), 54, 91; IIT (integrated information theory), 54, 91; LRTC (long-range temporal correlation), 17, 27–30; NCC (neural correlates of consciousness), 53; nestedness, 31–3, 91; neuronal specialness, 54–5, 99, 104–5; noise-like signals, 15; PLE (power law exponent), 27–30; resting state, 28, 51, 60–1; scale-free activity, 14–17, 26–35; self-similarity of world and brain, 31–3;

sensory functions, 37–8; temporal windows for world stimuli, 13–14, 19–20; timescales, 28–9, 31, 96–8; whole-brain organization, 18, 88, 91–3
construction view of time, 5–6, 12–13, 102
container view of time, 4–5, 12–13
Copernican revolution analogy, 102–5
cross-species relations. *See* non-human species
currency, time as common. *See* time, as common currency

Dainton, Barry, 4–5
dance. *See* music and dance
Darwin, Charles, 102–3, 104
default-mode network. *See* DMN (default-mode network)
de Pasquale, Francesco, 18–19
depression, 77–85, 99; bipolar disorder, 82; desynchronization of inner and outer time, 78, 80–1; duration of time, 84–5; frequencies (slow-fast), 77, 83, 84, 86, 99; horse racing analogy, 84; measurement of time speed, 79–81; neuroscience, 79–84, 99; perception of inner time speed, 77–8, 79, 84; self-focus vs outer focus, 82–4; sense of self, 81–5; time speed, 84–5, 99. *See also* mood disorders
Descartes, René, 53–4, 67–8, 99
DFA (detrended fluctuation analysis), 27–30
DMC (dual model of cognition), 51
DMN (default-mode network), 18–19; CMS (cortical midline structures), 60–3, 81–3; cross-network interaction, 18–20; in depression, 82–3; internal inputs, 51; music example, 92–4; neuroimaging, 18; resting state, 51, 82; self-reference, 82; spontaneous activity, 18–19, 22–3, 82; temporal integration, 92; temporal windows, 13–14, 19–20; timescales, 92, 98; uni-transmodal topography, 92, 98
dualism, mind-brain-body. *See* mind-body and world-brain problem
duration of time, 13–14, 62–7, 84; ACW (autocorrelation windows), 14, 24–5, 62, 66, 82, 92; brain's inner time, 13–14, 62–7, 70–3, 102–5; change and persistence, 59–60, 66–73; construction of inner time, 65–6, 82, 92, 102–5; as continuity of time, 13–14; defined, 60, 70, 84; in depression, 84; elapsed time, 70–3, 84; first- and third-person perspective, 36, 70–1, 95–6; philosophers on, 70–1; scale-free activity, 14–17, 62–3; sense of self, 62–73, 82; speed as velocity and duration, 13–14; spontaneous activity, 62–3; surfing analogy, 73; time of our inner subjective experience, lived time, 60; time paradox of self, 59–60, 72–3. *See also* self, neuroscience; speed of time

earth. *See* world
Edelman, Gerald, 56
EEG (electroencephalogram) studies, 18, 22, 27, 30, 44, 64, 66–7, 98
emotions: change and persistence, 58–60; in depression, 82; music example, 97; sense of self, 81–5; world-brain relation, 42, 47. *See also* mental features; mood disorders
entrainment, 45
environment. *See* world

fMRI studies, 18, 28, 43–4, 62–4, 66–7
fractals, 14, 15, 31–3. *See also* nestedness
Fuchs, Thomas, 78

GNWT (global neuronal workspace theory), 54, 91

INDEX

Goldstein, Kurt, 22, 23

Hardstone, Richard, 33
He, Biyu J., 16
heart. *See* body, heart
Heraclit, 8, 25
Hipp, Jeorg F., 18–19
Hippocrates, 4
horse racing analogy, 74–6, 84
Hua, Jingyu, 98
Huang, Zirui, 62, 63
Hume, David, 20–1, 50–1
Hurst exponent, 33

identity, 66–7. *See also* self
IIT (integrated information theory), 54, 91
illness. *See* mental illness and mental health
inner and outer time: brain's inner time as part of world's inner time, 9, 12–13, 24–5, 97, 102–5; construction view of time, 12–13, 102–5. *See also* brain, inner time; temporo-spatial theory of consciousness (TTC); time, as common currency; world, outer time

James, William, 9, 77

Kairos (moment in time), 3–5, 6
Kant, Immanuel, 21–2, 24, 50–2
ketamine. *See* consciousness, extended
Kolvoort, Ivar, 65

language analogy, 46–7
Lashley, Karl, 22
Lechinger, Julia, 44
LIS (locked-in syndrome), 37–40
Logothetis, Nikos, 88
LRTC (long-range temporal correlation), 15, 27; brain's inner time with world's outer time, 16–17; correlated brain and world, 33–5; DFA (detrended fluctuation analysis), 27–30; PLE (power law exponent), 27–30; scale-free activity, 15–17, 27–31; walking analogy, 33–5
LSD. *See* consciousness, extended

mania: common currency of time, 84–5, 99; perception of time speed, 77–81, 84–5. *See also* mood disorders
McGinn, Colin, 52–3
meditation, 33, 41
MEG (magnetoencephalography), 45, 48, 50
mental features, 7–10, 11, 99–101, 104–5; active vs passive models of brain, 21–5; artificial agents (AA), 10–11, 95–6; attention (on- and off-thoughts), 98; change and persistence, 8–9, 30–1, 56–7; common currency of time, 97; dynamic time as basis for, 7–10, 25, 102–5; key questions, 7, 8, 9, 11; LIS (locked-in syndrome), 37–40; mind-body and world-brain problem overview, 96–105; mind-brain-body dualism, 37, 40–2, 50, 52–4, 99–101; music example, 97; non-human species, 10; ocean waves analogy, 6–7, 9, 11; speed of time in brain, 76–7; timescales, 96–101; TTC (temporo-spatial theory of consciousness) overview, 35, 55–7; water states analogy, 8; world-brain relation, 7–10, 11, 16–17, 24–5, 27, 97, 99–105. *See also* consciousness; emotions; mind; mind-body and world-brain problem; self; temporo-spatial theory of consciousness (TTC)
mental illness and mental health: brain injuries, 39; common currency of time, 47; LIS (locked-in syndrome), 37–40; meditation, 33, 41; time speed

perception, 78, 81, 84–5; URWS (unresponsive wakefulness), 39; world-brain relation, 47. *See also* consciousness, loss or reduced; depression; mood disorders

mind, 7–10, 11, 53, 86–7, 100–1; active vs passive models, 20–5; brain-mind relation, 97–101; change and persistence, 8–9; common currency of time, 23–4, 25, 46–7, 55–7, 96–9; dynamic time as basis for, 6–10, 25, 97–105; key questions, 7, 8, 9; language analogy, 46–7; mind-body and world-brain problem overview, 96–105; neural correlates, 21, 48, 53; philosophers on, 20–2; TTC (temporo-spatial theory of consciousness) overview, 35, 55–7; world-brain relation, 7–10, 11, 16–17, 24–5, 96–101. *See also* consciousness; mental features; mind-body and world-brain problem; temporo-spatial theory of consciousness (TTC)

mind-body and world-brain problem, 11, 86–7, 96–105; common currency of time, 24, 25, 46–7, 96–105; mind as dynamic and temporal, 6–7, 102–5; mind-body problem, 86–7, 99–105; mind-brain-body dualism, 50, 52–4, 99–101; philosophers on, 11, 99; timescale stochastic matching, 93–6, 99–100; world-brain problem in place of mind-body problem, 11, 55–7, 99–105. *See also* temporo-spatial theory of consciousness (TTC)

moment in time (Kairos), 3–5, 6

mood disorders, 77–85, 99; bipolar disorder, 80, 82; common currency of time, 85, 99; desynchronization of inner and outer time, 78, 80–1; horse racing analogy, 84; mania, 77–81, 84–5, 99; measurement of time speed, 79–81; neuronal variability, 79–81, 84; neuroscience, 77–85, 99; perception of time speed, 77–81, 84–5; timescales and frequencies, 77, 84–5, 86, 99; world-brain relation, 77–81, 84–5. *See also* depression

motor functions, 37–40; coma and vegetative states, 38–9; consciousness, 37–40; dance example, 36–7; LIS (locked-in syndrome), 37–40; music example, 36–7, 97; neuroscience, 37–8; temporal windows, 13–14, 19–20; URWS (unresponsive wakefulness), 39; world-brain relation, 36–40

music and dance, 36–7; alignment of sensory and motor functions, 36–7; brain topography, 92–4; common currency of time, 97; dance, 36, 38, 94; duration of time, 70; emotions, 97; motor and sensory functions, 36–7; tango analogy, 27, 38, 42, 43, 44, 57; timescales, 92–4, 97; world-brain relation, 36–7, 42, 93–4

Nagel, Thomas, 52

NCC (neural correlates of consciousness), 21, 48, 50, 53

nestedness, 14–17, 31–4; brain's inner time and world's outer time, 16–17, 34, 40–1, 102–5; consciousness, 91; fractals, 14, 15, 31–3; frequencies (slow-fast), 15, 26; Roman cauliflower analogy, 32; Russian dolls analogy, 15, 16, 31, 32, 61–2; scale-free activity, 15, 16, 17, 26, 31–2; self, layers of, 61–2; self-similarity of world and brain, 31–3; TTC (temporo-spatial theory of consciousness), 91; world-brain relation, 16–17, 26, 31–5, 40–1. *See also* temporo-spatial theory of consciousness (TTC)

neural networks: convergence of temporal duration and spatial extension in brain, 18–20; cross-network inter-

action, 18–20; spontaneous activity, 18. See also DMN (default-mode network); spontaneous activity in brain neuroscience: active vs passive models of brain, 20–5; artificial intelligence models, 90–1; Hurst exponent, 33; internal vs external contents of consciousness, 51; mind-brain-body dualism, 50, 52–4, 99–101; neural correlates, 21, 48, 50, 53; neuroimaging, 18; neuronal specialness, 54–5, 99, 104–5; self as temporal, 68–9; self-similarity of world and brain, 31–3; time as change and persistence, 8–9, 13; views on time, 12–13. See also brain, neuroscience; consciousness, neuroscience; DMN (default-mode network); nestedness; neural networks; scale-free activity; scale-free activity, neuroscience; self, neuroscience; spontaneous activity in brain, neuroscience
non-human entities. See artificial intelligence (AI) and artificial agents (AA)
non-human species, 10, 87–8, 100–1; bats, 52, 88–9, 94, 100; common currency of time, 87–9, 96–7; mental features, 10; pets, 87–8; timescale alignment, 87–9, 94, 100–1; world-brain relation, 10, 87–9

ocean wave analogy, 6–7, 9, 41–2, 63, 73. See also surfing analogy
other-than-human species. See non-human species
outer time. See world, outer time

PACC (perigenual anterior cingulate cortex), 48–9, 50, 60–5
Park, Hyeong-Dong, 48
PCC (posterior cingulate cortex), 60–5
perception: of inner and outer time, 77–81; neural correlates, 21; neuronal variability, 79–81; passive model of the brain, 20–1; synchronization of inner and outer time, 78, 80–1; of time speed in mood disorders, 77–81. See also speed of time
perspective (point of view), 36, 50, 52–3, 70–1, 95–6
philosophy and philosophers: artificial intelligence, 90–1; Bergson, 5, 70–1, 73; brain models, active vs passive, 20–5; consciousness, 50–4; Descartes, 53–4, 67–8, 99; duration of self, 70–1; Eastern philosophers, 6–7; Hume, 20–1, 50–1; Kant, 21–2, 24, 50–2; self, 59, 67–73
philosophy and time: as change and persistence, 5–6, 8–9, 25, 30–1, 56–7, 59–60; as construction, 5–6, 12–13, 102; as container, 4–5, 12–13; as dynamic, 5–6, 25, 102–5; mind-body and world-brain problem overview, 96–105; relationism, 5–6; self as atemporal vs temporal, 67–73; time as basic to world, 40–1, 55–7, 100–5. See also mind-body and world-brain problem; time
physics and time: as basic to world, 40–1, 55–7; brain and time as separate vs convergent, 24; as change and persistence, 5, 13; as construction, 5–6, 12–13, 102; as a container for moments, 4–5, 12–13; contemporary vs classical physics, 5–6; as dynamic, 5, 25, 102–5
PLE (power law exponent), 16–17, 27–30, 32, 62, 63–4, 65, 67
Prescott, Tony, 90, 95, 96
psychostimulants, 27, 29, 34, 35, 41. See also consciousness, extended

Qin, P., 61
quantum theory of time, 13

Raichle, Marcus, 22–3, 24
relationism and time, 5–6
research. *See* neuroscience; philosophy and philosophers
Richter, Craig G., 45
river or stream analogy, 8, 9, 31, 77–8
Roman cauliflower analogy, 32
Rostami, Samira, 98
Russian dolls analogy, 15, 16, 31, 32, 61–2

Scalabrini, Andrea, 83
scale-free activity, 14–17, 26–33; consciousness, 26–33; defined, 31; mental features, 16; Roman cauliflower analogy, 32; Russian dolls analogy, 15, 16, 31, 32, 61–2; self, 62–4, 66–7; self-similarity of world and brain, 31–3; surfing analogy, 35; as ubiquitous, 17; world-brain relation, 31–5; world's outer time, 16–17
scale-free activity, neuroscience, 14–17, 26–35, 62–7; active construction of inner temporal duration, 24–5, 102; CFC (cross-frequency coupling), 15, 45, 62, 82; DFA (detrended fluctuation analysis), 27–30; fMRI studies, 28; frequencies (slow-fast), 14–15, 16, 26, 27–30; Hurst exponent, 33; LRTC (long-range temporal correlation), 15, 17, 27, 31; nestedness, 15, 16, 17, 26, 31–2; noise-like signals, 15; PLE (power law exponent), 27–30, 62, 63–4, 65, 67; resting state, 28; spontaneous activity, 14–18, 24–5, 26, 28, 62–3; temporal duration, 62; timescales, 16, 26, 27, 28, 31
science: Copernican revolution analogy, 102–5. *See also* neuroscience; physics and time
self, 7–11, 58–73, 81–5; artificial agents (AA), 10–11, 90, 95–6; brain-body coupling, 50; brain's inner time, 60–9, 72–3, 76, 102–5; change and persistence, 8–9, 11, 30–1, 56–60, 66–73; cognitive functions, 81; common currency of time, 65–6, 68–9, 96–9; consciousness, internal contents, 47–51, 57; in depression, 81–5; duration of self, 62–7, 69–73; elapsed time, 70–3, 74, 84; first- and third-person perspective, 36, 50, 52–3, 70–1, 73, 95–6; "I" and "me," 50; identity, 67–8; key questions, 7, 8, 9; mental features, 81; philosophers on, 59, 67–73; self-consciousness, 61, 63–4, 67, 81–2; self-reference effect, 81–2; self-similarity of world and brain, 31–3; sense of self, 64–7, 81–5; temporal vs atemporal, 67–73; terminology, 81; time as basis for, 7–10; time paradox, 59–60, 72–3; world-brain relation, 11, 24–5. *See also* consciousness; emotions; mental features; perception
self, analogies: ocean waves, 63, 73; river, 8, 9, 31; Russian dolls, 61–2; surfer, 11, 58, 73
self, neuroscience, 60–9; ACW (autocorrelation windows), 14, 24–5, 62, 66, 82, 92; brain topography, 49, 57, 61–2, 91–3; CMS (cortical midline structures), 60–5, 81–3; DMN (default-mode network), 82–3, 98; EEG studies, 64, 66–7; fMRI studies, 62–4, 66–7; frequencies (slow-fast), 62–9, 72; layers (mental, interoceptive, exteroceptive), 49, 57, 61–2; nestedness, 61–2; noise-like signals, 15; PACC (perigenual anterior cingulate cortex), 60–5; PCC (posterior cingulate cortex), 60–5; PLE (power law exponent), 27–30, 62–5, 67; prefrontal cortex (VMPFC and DMPFC), 49, 50, 82, 98; resting state, 60–1, 66, 82; scale-free activity, 14–17, 63–4, 66–7; self-consciousness scale

(SCS), 63–4; spatial and temporal structure, 60–7; spontaneous activity, 62–3, 65, 82; timescales, 66–9, 72–3, 96–9

sensory functions, 36–40; coma and vegetative states, 38–9; consciousness, 37–40; external contents of consciousness, 51; LIS (locked-in syndrome), 37–40; music example, 36–7, 97; neuroscience, 37–8; temporal windows, 13–14, 19–20; URWS (unresponsive wakefulness), 39; world-brain relation, 36–40, 57. *See also* body

sequential time (Chronos), 3–5, 6

Sherrington, Charles, 20, 22

Shinomoto, Shigeru, 87–8

Singer, Wolf, 88

sleep: brain's spontaneous activity, 44; clouded consciousness, 29, 30; deprivation, 27–8, 30; frequencies (slow-fast), 29, 30; heart rate, 44; LIS (locked-in syndrome), 37–40; loss of sense of self, 98–9; sleep stages (N1 to N3, REM), 29, 30, 44, 99. *See also* consciousness, loss or reduced

Smith, David, 66

SMN (somatomotor network), 79–81

Smolin, Lee, 5

soccer players analogy, 77

soup analogy, 28–9

speed of time, 74–80, 84–5; common currency of time, 85; defined, 74; fast vs slow external reactions, 76–7; horse racing analogy, 74–6, 84; measurement of, 74, 79–81; mental features, 76–7, 97; mental speed, 76–7; as objective quantity, 74; perception of time, 77–81; psychiatrists on, 78; soccer players analogy, 77; stream of consciousness, 9–10, 77–8; synchronization of inner and outer time, 78, 80–1. *See also* duration of time; mood disorders

speed of time, neuroscience: CMS (cortical midline structures), 81–3; co-occurring speeds, 75–6; DMN (default-mode network), 77, 82–3; frequencies (slow-fast), 75–7, 82–4; neural timing circuit, 79; neuronal speed, 76–7; neuronal variability, 79–81, 84; prefrontal cortex (VMPFC and DMPFC), 82; resting state, 82, 83; SMN (somatomotor network), 79–81; spontaneous activity, 75, 82; timescales, 75–7; velocity, 10, 74, 75, 84; VN (visual network), 79–81. *See also* duration of time

spontaneous activity in brain, 14–25; active vs passive models of brain, 20–5; consciousness shaping by, 49, 50, 51–2; heart-brain relation, 44, 45–6, 48–9; horse racing analogy, 74–6, 84; inner time, 62–3; neuroimaging, 18; philosophers on, 20–2; self, 62–4, 65, 82; sleep, 44; speed of time, 75–7; stomach-brain relation, 42, 45–7; world-brain relation, 16–17, 24–5

spontaneous activity in brain, neuroscience, 17–19, 20–5; ACW (autocorrelation windows), 14, 24–5, 66, 82, 92; CFC (cross-frequency coupling), 15, 45, 62, 82; co-occurring speeds, 75–6; DMN (default-mode network), 18–19, 22–3; EEG studies, 22, 44; frequencies (slow-fast), 14–15, 75–7; heart rate, 44; links between spatial and temporal dimensions, 19; nestedness, 15; neural correlates, 50; neural networks, 18; phase shifting or locking, 44; resting state, 18, 60–1, 82; scale-free activity, 14–18, 24–5, 26, 28, 62–3; self-consciousness scale (SCS), 63–4; spatial structure, 17–18, 19; temporal structure, 18, 19, 62–3; timescales, 75; whole-brain organization, 18, 88, 91–3

stomach-brain relation, 42, 45–7

stream of consciousness, 9–10, 77–8. *See also* consciousness
surfing analogy, 10, 11, 35, 41–3, 58, 73

Tallon-Baudry, Catherine, 48, 50
tango analogy, 27, 38, 42, 43, 44, 57
temporal duration. *See* duration of time
temporo-spatial theory of consciousness (TTC), 35, 55–7; brain-heart relation, 44; common currency of time, 24, 25, 55–7, 96–101; consciousness as non-special, 54–7, 99; consciousness as world-brain alignment, 35, 40–1, 43, 56–7; mind-body and world-brain problem overview, 96–105; nestedness, 91; world-brain relation, 7–10, 11, 16–17, 35, 102–5. *See also* mind-body and world-brain problem; time, as common currency
time, 3, 6–10, 40–1, 55–7, 100–1; as basic to world, 3, 40–1, 55–7, 100–5; change and persistence, 8–9, 13, 25, 30–1, 56–7, 58–60; Chronos (continuum of time), 3–7; construction view, 5–6, 12–13, 102; container view, 4–5, 12–13; Copernican revolution analogy, 102–5; dynamic view, 6–10, 97–105; as inner time, 13, 102–5; Kairos (moment in time), 3–5, 6; key questions, 7, 8, 9; neuroscience view, 13; as outer time, 12–13; quantum theory, 13; relationism, 5–6, 13; and space, 17–20; static vs dynamic view, 25. *See also* brain, inner time; duration of time; mind-body and world-brain problem; philosophy and time; physics and time; speed of time; world, outer time
time, as common currency, 8–10, 46–7, 55–7, 96–105; body-brain relation, 46–7, 97; between brain, consciousness, and world, 26–7, 84–5; brain's inner time as part of world's inner time, 9, 12–13, 24–5, 97, 102–5; from brain to consciousness, 26–7, 30–1, 40–1, 55–7; from brain to mind, 23–4, 25, 97; duration of self, 65–6; EEG studies, 98; language analogy, 46–7; mental features, 47, 97, 104; mind-body and world-brain problem overview, 24, 25, 96–105; in mood disorders, 84–5; time as basic to world, 3, 40–1, 55–7, 97–101; timescales, 87–9, 97–8; TTC (temporo-spatial theory of consciousness) overview, 35, 55–7; world-brain relation, 7–10, 11, 24–5, 102–5. *See also* mind-body and world-brain problem; temporo-spatial theory of consciousness (TTC)
time, analogies: horse racing, 74–6, 84; ocean waves, 6–7, 9, 41–2, 63, 73; river or stream, 8, 9, 31, 77–8; soccer players, 77; sunset, 25; surfing, 11, 35, 41–3, 73; tango, 27, 42, 57; trash can, 5; walking, 33–5
trans-species relations, 87–9. *See also* non-human species
trash can analogy, 5
TTC. *See* temporo-spatial theory of consciousness (TTC)

URWS (unresponsive wakefulness), 39

vegetative states, 38–9. *See also* consciousness, loss or reduced
VN (visual network), 79–81

walking analogy, 33–5
water states analogy, 8
Whitehead, Alfred North, 5–6
Wolff, Annemarie, 64
world: consciousness based in, 26–7, 39–41; mind-body and world-brain problem overview, 96–105; surfing analogy, 11, 35, 41–3, 73; TTC (temporo-spatial

theory of consciousness) overview, 35, 55–7; world-brain relation, 7–10, 11, 16–17, 24–5, 27; world's inner time, 16–17, 27, 42, 102–5. *See also* mind-body and world-brain problem; temporospatial theory of consciousness (TTC)

world, outer time, 16–17, 26–7, 55–7, 102; active vs passive models of brain, 20–5; brain's inner time as part of, 9, 12–13, 24–5, 102–5; brainwave-worldwave relation, 38, 42–3; consciousness, 17, 26–7, 33–5; as a container, 4–5, 12–13; correlated brain and world, 33–5, 38; as dynamic, 7, 102–5; fluctuations in timescales, 89, 97–8; frequencies and fluctuations, 16, 26; LRTC (long-range temporal correlation), 17, 33–4; mind alignment of inner and outer timescales, 96–101; mind-body and world-brain problem overview, 96–105; nestedness, 16–17, 26, 31–3, 34, 40–1; non-human alignment, 89; perception of inner and outer time, 77–81; PLE (power law exponent), 16–17; scale-free activity, 16–17, 24–5, 26, 31–5; self-similarity of world and brain, 31–3; synchronization of inner and outer time, 78, 80–1; temporal windows for brain stimuli, 13–14, 19–20; timescales, 16, 26, 87–9, 96–9; timescale stochastic matching, 93–6, 99–100; TTC (temporo-spatial theory of consciousness), 35, 55–7; world-brain relation, 7–10, 11, 14, 16–17, 24–5, 31–5, 38, 40–3, 55–7, 96–105. *See also* brain, inner time; mind-body and world-brain problem; temporo-spatial theory of consciousness (TTC); time, as common currency

world, outer time, analogies and examples: seismic earth waves, 16–17, 26, 89; surfing analogy, 11, 35, 41–3, 73; tango analogy, 27, 42, 57; ultrasonic waves, 89, 94, 100; walking analogy, 33–5; water states analogy, 8

Zilio, Federico, 38